Benjamin Huh's ACT® MATHEMATICS

Benjamin Huh, M.Ed.

University of California - Berkeley / Applied Math, BA

University of Washington / Math Education, M Ed.

Director / Berkeley Prep & Consulting, Seattle, WA

About the Author

The Author, Benjamin Huh has B.A. degree in Applied Mathematics from UC Berkeley and M.Ed degree in Mathematics Education from University of Washington. He has been teaching secondary and college mathematics for more than 20 years, including algebra I, algebra II, geometry, pre-calculus, AP calculus AB & BC, AP Statistics as well as advanced math in multi variable calculus, differential equation and linear algebra. He has been director of Berkeley Learning Center for more than decades and taught many High school students in the U.S. and Korea helping them not only in the subject of Mathematics producing many perfect scored students in SAT I and II math, but also in consulting and counselling of the US College Admission and Application for his students to gain admissions from the highly selective colleges and universities in the nation including IVY League schools and top tier schools.

Benjamin Huh

University of Washington, M.Ed.
University of California - Berkeley, Mathematics BA

- 2015-Present Berkeley Prep & Consulting - Bellevue, WA, Director/Math Instructor
- 2009-2014 Berkeley Learning Center - Irvine, CA, Director/Math Instructor
- 2004-2008 Berkeley Learning Center - Gangnam/Seoul, Korea, Director/Math Instructor
- 1999-2004 Berkeley Learning Center - Seattle/Tacoma, WA, Director/Math Instructor

Acknowledgments

I thank my friend, Andrew An, for his work on editing, formatting and the preparation of the manuscript of this book.

ISBN-13: 978-0996564717
ISBN-10: 0996564713

Contents

PART 2
MODEL TESTS

Contents

PART 2
MODEL TESTS

1. ACT - Math Contents*

60 multiple - choice questions and topics covered - as follow;

Content	Approximate Number of Questions
Pre-Algebra/Elementary Algebra	**21-27**
Pre-Algebra. Questions in this content area are based on basic operations using whole numbers, decimals, fractions, and integers; place value; square roots and approximations; the concept of exponents; scientific notation; factors; ratio, proportion, and percent; linear equations in one variable; absolute value and ordering numbers by value; elementary counting techniques and simple probability; data collection, representation, and interpretation; and understanding simple descriptive statistics.	12-15
Elementary Algebra. Questions in this content area are based on properties of exponents and square roots, evaluation of algebraic expressions through substitution, using variables to express functional relationships, understanding algebraic operations, and the solution of quadratic equations by factoring.	9-12
Intermediate Algebra/Coordinate Geometry	**18-24**
Intermediate Algebra. Questions in this content area are based on an understanding of the quadratic formula, rational and radical expressions, absolute value equations and inequalities, sequences and patterns, systems of equations, quadratic inequalities, functions, modeling, matrices, roots of polynomials, and complex numbers.	9-12
Coordinate Geometry. Questions in this content area are based on graphing and the relations between equations and graphs, including points, lines, polynomials, circles, and other curves; graphing inequalities; slope; parallel and perpendicular lines; distance; midpoints; and conics.	9-12
Plane Geometry/Trigonometry	**15-21**
Plane Geometry. Questions in this content area are based on the properties and relations of plane figures, including angles and relations among perpendicular and parallel lines; properties of circles, triangles, rectangles, parallelograms, and trapezoids; transformations; the concept of proof and proof techniques; volume; and applications of geometry to three dimensions.	12-15
Trigonometry. Questions in this content area are based on understanding trigonometric relations in right triangles; values and properties of trigonometric functions; graphing trigonometric functions; modeling using trigonometric functions; use of trigonometric identities; and solving trigonometric equations.	3-6

(* From ACT – Official Site)

2. Test Day Checklist**

- Report to your assigned test center by the Reporting Time (8:00 a.m.) listed on your ticket. You will NOT be admitted to test if you are late.
- Bring a printed copy of your ticket to the test center. You will not be admitted to test if you do not have a printed copy of your ticket.
- Bring acceptable photo identification. You will not be admitted to test if your ID does not meet ACT's requirements.
- Testing staff will check your photo ID and ticket, admit you to your test room, direct you to a seat, and provide test materials.
- Bring sharpened soft lead No. 2 pencils with good erasers (no mechanical pencils or ink pens). Do not bring any other writing instruments; you will not be allowed to use them.
- Be ready to begin testing after all examinees resent at 8:00 a.m. are checked in and seated.

3. Tips for Taking the ACT

- Use a soft lead No. 2 pencil wit good eraser. Do not use a mechanical pencil or ink pen; if you do, your answer document cannot be scored accurately.
- Answer the easy questions first, then go back and answer the more difficult ones if you have time remaining on that test.
- On difficult questions, eliminate as many incorrect answers as you can, then make an educated guess among those remaining.
- Answer every question. Your scores on the multiple-choice tests are based on the number of questions you answer correctly. There is no penalty for guessing.
- If you complete a test before time is called, recheck your work on that test.
- Mark your answers properly. Erase any mark completely and cleanly without smudging.

4. Tips for the ACT Multiple-choice Tests

- The ACT-Mathematics is a 60 questions - 60 minutes test with no penalty of guessing.
- Read each question carefully to make sure you understand the type of answer required.
- If you choose to use a calculator, be sure it is permitted, is working on test day, and has reliable batteries. Use your calculator wisely.
- Solve the problem.
- Locate your solutions among the answer choices.
- Make sure your answer the question asked.
- Make sure your answer is reasonable.
- Check your work.

(** From ACT – Official Site)

ACT Math Reference – Formula and Notes

1. Quadratic Function,

Vertex Form:

$y = ax^2 + bx + c = a(x - h)^2 + k$

Vertex (h, k), where $x = \frac{-b}{2a}$, or called a line of symmetry, and $V = (x, f(x))$

(eg. 1) What is the vertex of $y = f(x) = x^2 - 3x + 2$?

$y = f(x) = x^2 - 3x + 2$ has vertex at $x = \frac{-b}{2a}$.

$\therefore x = \frac{-(-3)}{2} = 1.5$

$\therefore f(1.5) = (1.5)^2 - 3(1.5) + 2 = 2.25 - 4.5 + 2 = -0.25$

$\therefore$ vertex, $v = (1.5, -0.25)$

Quadratic Equation:

Given $ax^2 + bx + c = 0$, $x = \frac{-b \pm \sqrt{b^2 - 4ac}}{2a}$

And suppose its zeros are x_1, and x_2, then, the sum of $x_1 + x_2 = \frac{-b}{a}$ and the product $x_1 \cdot x_2 = \frac{c}{a}$.

(eg. 1) In the equation $x^2 + kx + 48 = 0$, one root is twice the other root. The value(s) of k is (are)?

To solve this problem, we need to refer to the following formula from the quadratic eq.:
Given $ax^2 + bx + c = 0$, suppose its zeros are x_1, and x_2.

Then, the sum of $x_1 + x_2 = \frac{-b}{a}$, and the product $x_1 \times x_2 = \frac{c}{a}$.

Now, from our equation, $x^2 + kx + 48 = 0$, we get $a = 1$, $b = k$ and $c = 48$.
Also, the problem states that one root is twice of the other root, this, $x_2 = 2x_1$.

Now, the sum $x_1 + x_2 = x_1 + (2x_1) = 3x_1 = \frac{-b}{a}$, or $3x_1 = \frac{-k}{1} = (-)k$, or $x_1 = \frac{-k}{3}$. --- eq.(1)

Also, the product $x_1 \times x_2 = x_1 \times (2x_1) = 2x_1^2 = \frac{c}{a}$, or $2x_1^2 = \frac{48}{1} = 48$, or $x_1^2 = 24$. --- eq.(2)

Therefore, from the above eq.(1) and (2), we get $x_1 = \pm 2\sqrt{6} = \frac{-k}{3}$, and thus, $k = \pm 6\sqrt{6}$.

(eg. 2) If the roots of the equation $ax^2 + bx + c = 0$ are α and β, the value of $\frac{(\alpha + \beta)^2}{\alpha\beta}$ in terms of a, b and c is?

Recall from the properties of the quadratic function, we have sum of $\alpha + \beta = \frac{-b}{a}$, $\alpha\beta = \frac{c}{a}$.

$$\therefore \frac{(\alpha + \beta)^2}{\alpha\beta} = \frac{\left(\frac{-b}{a}\right)^2}{\frac{c}{a}} = \left(\frac{-b}{a}\right)^2 \cdot \frac{a}{c} = \frac{b^2}{a^2} \cdot \frac{a}{c} = \frac{b^2}{ac}$$

Discriminant Testing:

For $D = b^2 - 4ac$,

- $D > 0$, then there are two real roots;
- $D = 0$, then one double roots;
- $D < 0$, there is no real roots.

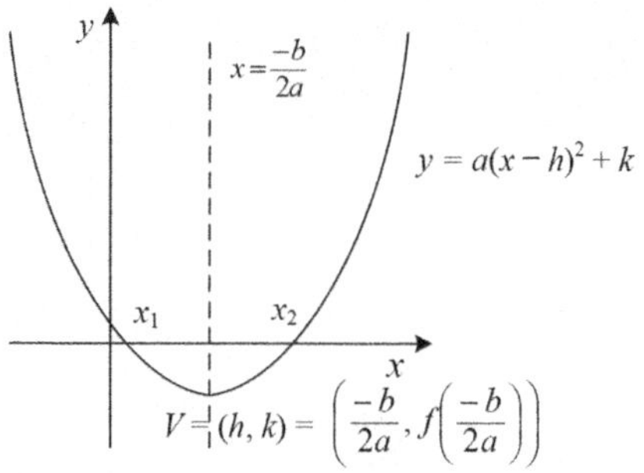

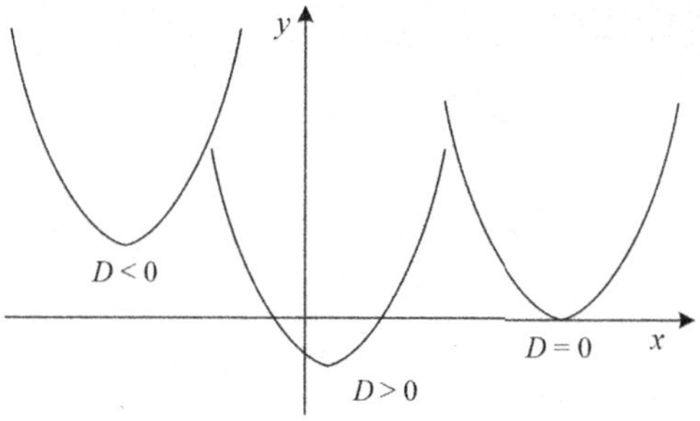

Note: **Cubic function;**

Let $f(x) = ax^3 + bx^2 + cx + d$, then sum of the three roots, $x_1 + x_2 + x_3 = -\frac{b}{a}$

and the product of the roots, $x_1 \cdot x_2 \cdot x_3 = -\frac{d}{a}$.

(eg. 1) If the roots of the equation $x^3 + px^2 + qx + r = 0$ are s_1, s_2 and s_3, then $s_1^2 + s_2^2 + s_3^2 = ?$

Recall from the cubic equation, $ax^3 + bx^2 + cx + d = 0$, the sum of 3 roots $x_1 + x_2 + x_3 = -\frac{b}{a}$,

$x_1x_2 + x_2x_3 + x_3x_1 = \frac{c}{a}$, $x_1 \cdot x_2 \cdot x_3 = -\frac{d}{a}$. In this problem, $a = 1$, $b = p$, $c = q$, $d = r$.

$\therefore s_1 + s_2 + s_3 = -p$, $s_1s_2 + s_2s_3 + s_3s_1 = q$

$\therefore s_1^2 + s_2^2 + s_3^2 = (s_1 + s_2 + s_3)^2 - 2(s_1s_2 + s_2s_3 + s_3s_1) = (-p)^2 - 2q = p^2 - 2q$.

2. Shifting and Translating of The Functions;

Let $y = f(x - h) + k$, represents the standard form of $f(x)$, then $f(x)$ has a horizontal translation of h unit, and a vertical shift of k unit. When $h > 0$, then move to the right, and $h < 0$, move to the left etc…

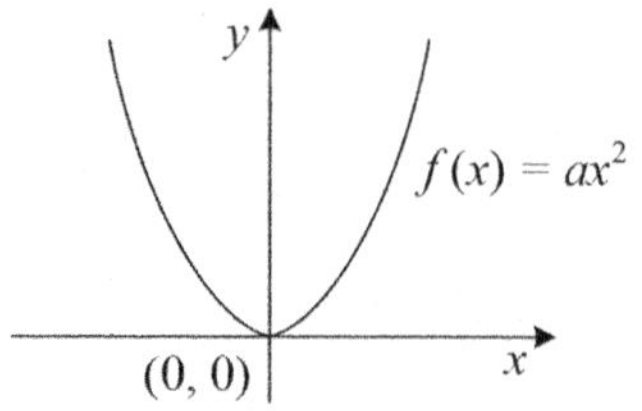

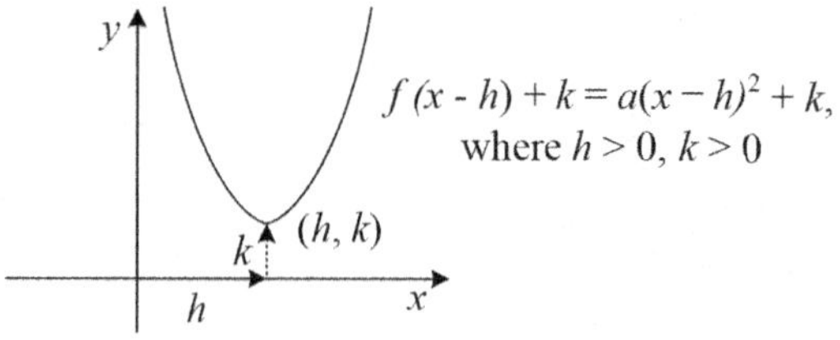

(eg. 1) Suppose the graph of $f(x) = x^2 + x + 1$ is translated 2 units up and 1 units right. If the resulting graph represents the graph of $g(x)$, what is the value of $g(-1.5)$?

$y = f(x - h) + k$, represents the standard form in which $f(x)$ has a horizontal translating of h unit, and a vertical shifting of k unit. When $h > 0$, then move to the right, and $h < 0$, move to the left etc…
Now, $g(x)$ represents the graph of $f(x)$ with $h = 1$, and $k = 2$.
Therefore, $g(x) = f(x - 1) + 2$, or plugging in $(x - 1)$ for x, and adding 2, we get $g(x) = [(x-1)^2 + (x-1) + 1] + 2 = (x-1)^2 + (x-1) + 3$. Thus, $g(-1.5) = [(-2.5)^2 + (-2.5) + 3] = 6.75$.

(eg. 2) Suppose the graph of $f(x) = x^2 - 4x + 7$ is translated 2 units left and 3 unit down. If the resulting graph represents $g(x)$, what is the value of $g(-0.5)$?

Since x is translated 2 units left, we get, $x \rightarrow x - (-2) = x + 2$. Also, y is 3 down, $y \rightarrow y - (-3) = y + 3$
$\therefore g(x) + 3 = f(x + 2)$, or $g(x) = f(x + 2) - 3 = (x + 2)^2 - 4(x + 2) + 7 - 3$

Now, $g\left(-\frac{1}{2}\right) = \left(\frac{3}{2}\right)^2 - 4 \cdot \frac{3}{2} + 4 = \frac{9}{4} - 6 + 4 = \frac{1}{4}$

ACT Math Reference – Formula and Notes

3. Polynomial and Rational Functions

A) The definition of even and odd function,

Even Function, if $f(-x) = f(x)$ or symmetric to y-axis, and
Odd Function, $f(-x) = -f(x)$ or symmetric to origin (0, 0).

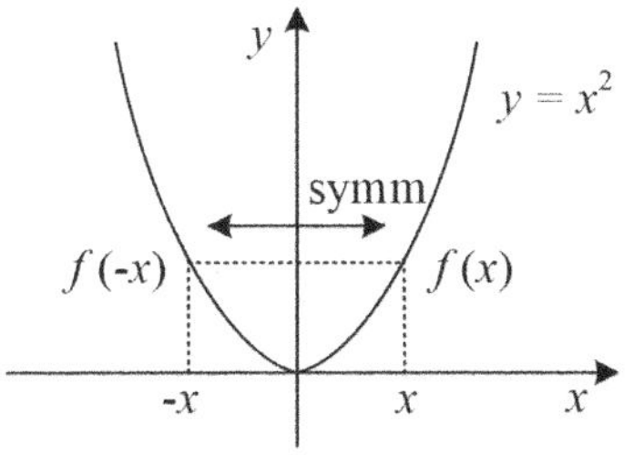

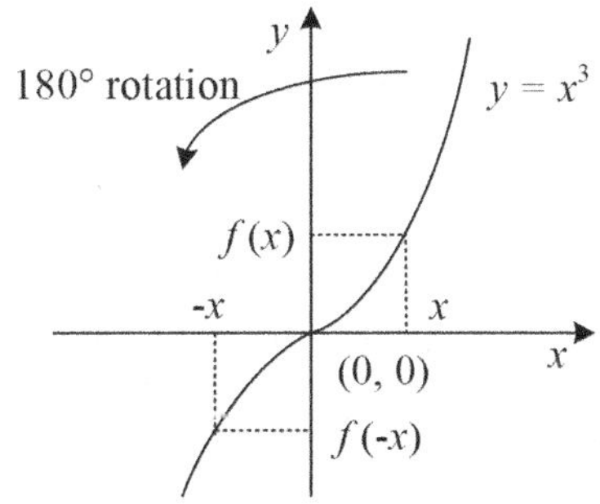

Even Function: $f(-x) = f(x)$
Even Function: symmetric to y-axis

Odd Function: $f(-x) = -f(x)$
Odd Function: symmetric to origin

(eg. 1) Which of the following is *not* an even function?

(A) $y = |x| + 1$ (B) $y = \cos x$ (C) $y = \log x^2$ (D) $y = x^2 + \tan x$ (E) $y = 4x^4 + 2x^2 + 3$

The definition of even function is, $f(-x) = f(x)$, or symmetric to y-axis. In this problem, instead of using the definition, it would be better to use a graphing calculator to see if any of the function given is not symmetric to y-axis.
The choice (D) $y = x^2 + \tan x$ is not symmetric to y-axis.

(eg. 2) If $f(x) = \sec x$ and $g(x) = 3x - 1$, which of the following is an even function (are even functions)?

I. $f(x)\cdot g(x)$
II. $f(g(x))$
III. $g(f(x))$

Since even function is symmetric to y-axis, using graphing utility,
let $f(x)g(x)$ as $y_1 = (\sec x)(3x - 1) = \frac{1}{\cos x}\cdot(3x - 1)$, and $f(g(x))$ as, $y_2 = \sec(3x - 1) = \frac{1}{\cos(3x - 1)}$.
Also, $g(f(x))$ as, $y_3 = 3(\sec x) - 1 = \frac{3}{\cos x} - 1$.
Then, we find that (III) $g(f(x)) = 3(\sec x) - 1$ is the only one that is symmetric to y-axis, thus it is even function.

(eg. 3) If $f(x) = 2x$ and $g(x) = x^2 + 1$, which of the following must be true?

I. $f(x)g(x)$ is an odd function.
II. $f(g(x))$ is an even function.
III. $g(f(x))$ is an even function.

To be even function, it is symmetric to y-axis, and also to be odd function, it should be symmetric to origin (0, 0). Now, by trying graphing utility for (I), (II) and (III),

- $y_1 = f(x)g(x) = 2x(x^2 + 1)$ shows symmetric to (0, 0) ∴ True, odd function.
- (II) $y_2 = f(g(x)) = 2(x^2 + 1)$ shows symmetric to y-axis. ∴ True, even function.
- (III) $y_3 = g(5(x)) = (2x)^2 + 1$ shows symmetric to y-axis. ∴ True, even function.

4. Exponential Functions

Problems involve such as calculating interest rate, population growth or any exponential growth or decay, we have the following formula in two different cases.

Discrete case: $A = A_0(1 + \frac{r}{n})^{nt}$, where r is annual rate, A_0 is an initial amount,

- n is number of times that annual rate r is compounded, and t is time period.
- Continuous case: $A = A_0e^{rt}$, or $C = C_0e^{kt}$, where positive k is for growth, and negative k for decay.

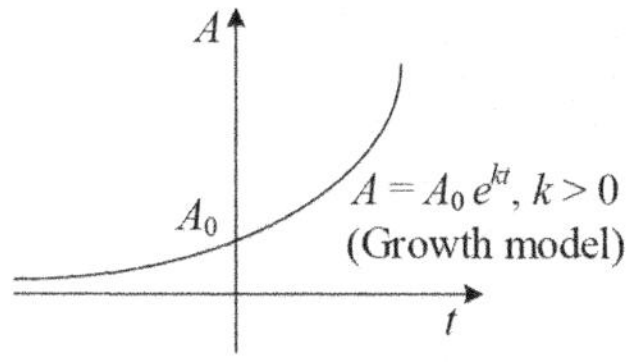

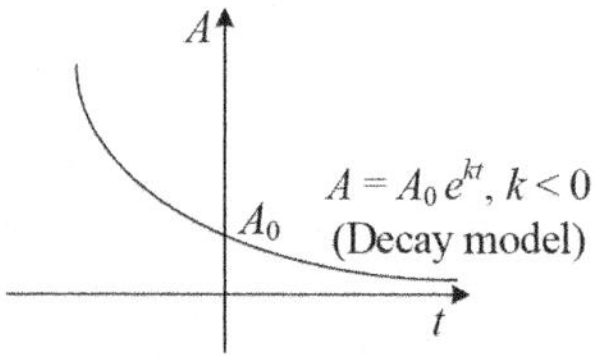

(eg. 1) If an initial investment of \$2,000 in a savings account is compounded continuously at an annual rate of 6 percent, how many years will it take the investment to be worth approximately \$5,000?

Notice that the saving is compounded continuously at $r = 6\%$ or, $r = 0.06$ annually.
In this problem, we choose the continuous case, because it is compounded continuously. $y = A_0e^{rt}$ where $A_0 = 2000$, $r = 0.06$, $y = 5{,}000$. This leads us to our eq. $5000 = 2000 \times e^{0.06t}$, or $\frac{5000}{2000} = e^{0.06t}$.

Applying natural log, ln on both sides, we get $\ln\frac{5}{2} = 0.06t$. Therefore, $t = \frac{\ln\frac{5}{2}}{0.06} = 15.3$

(eg. 2) When a certain radioactive element decays, the amount that exists at any time t can be calculated by the function $A(t) = A_0e^{\frac{-t}{1000}}$, where A_0 is the initial amount and t is the elapsed time in years. How many years would it take for an initial amount of 800 milligrams of this element to decay to 200 milligrams?

$200 = 800 \cdot e^{\frac{-t}{1000}}$ $\therefore \frac{200}{800} = \frac{1}{4} = e^{\frac{-t}{1000}}$

Now, applying "ln" on both sides, we get $\ln\frac{1}{4} = \ln e^{\frac{-t}{1000}} = \frac{-t}{1000}(\ln e) = \frac{-t}{1000}(1)$. $\therefore t = -1000 \cdot \ln\frac{1}{4} = 1{,}386$

5. The inverse of a function is defined by $f^{-1}(x)$, for which,

- $f(x)$ must be one to one function.
- $f(x)$ and $f^{-1}(x)$ are symmetric to $y = x$ line
- To get $f^{-1}(x)$, we need to switch x into y, and y into x.
- The Domain and Range of $f(x)$ become the Range and the Domain of $f^{-1}(x)$ by interchanging respectively.

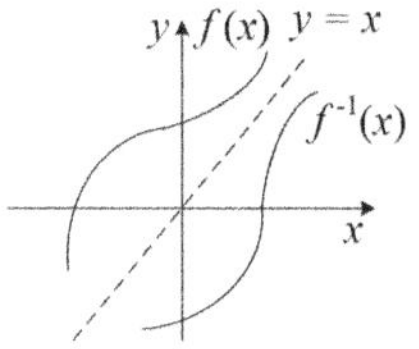

(eg. 1) What is the inverse of the function of $f(x) = \frac{1}{\sqrt{x-1}}$, $x > 1, y > 0$?

Here, $f(x) = \frac{1}{\sqrt{x-1}} = y$. Now, by interchanging x and y, we get $\frac{1}{\sqrt{y-1}} = x$ or $\frac{1}{\sqrt{(y-1)^2}} = x^2$.

That is $\frac{1}{y-1} = x^2$ or $y - 1 = \frac{1}{x^2}$, which results in $y = \frac{1}{x^2} + 1 = \frac{1+x^2}{x^2}$ with $x > 0, y > 1$.

6. Properties of log function:

1) $\log x \cdot y = \log x + \log y$
2) $\log \frac{x}{y} = \log x - \log y$
3) $\log x^m = m \cdot \log x$
4) $\log_x y = \frac{\log y}{\log x} = \frac{\ln y}{\ln x}$
5) $\log_x x = 1$, $\log_x 1 = 0$
6) $y = \log_a x$, if and only if $x = a^y$

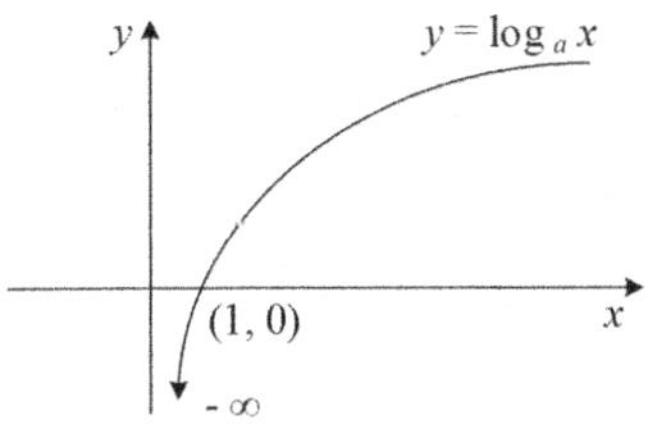

(eg. 1) If $\log_a 9 = 2$, then what is a?

$\log_a 9 = 2 \rightarrow 9 = a^2$ by the definition of logarithm. $\therefore a = 3$.

(eg. 2) Approximate $\log_8 13$?

$\log_A B = \frac{\log B}{\log A}$ $\therefore \log_8 13 = \frac{\log 13}{\log 8} = 1.23$

(eg. 3) If $\log_9 p = 3$ and $\log_3 q = 4$, p expressed in terms of q is?

$\log_9 p = 3 \rightarrow p = 9^3 = (3^2)^3 = 3^6$, $\log_3 q = 4 \rightarrow q = 3^4$

$\therefore p = 3^6 = (3^4)^{\frac{3}{2}} = (q)^{\frac{3}{2}}$

7. The formula for the distance *D* between a point (*x*, *y*) and a line *Ax* + *By* + *C* = 0;

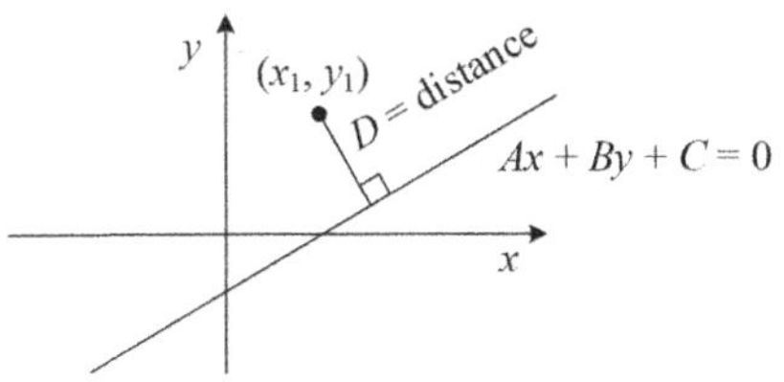

$$D = \frac{|Ax_1 + By_1 + C|}{\sqrt{A^2 + B^2}}$$

(eg. 1) If point $P(k, 2k)$ is 3 units from line $3x + 4y = 7$, then k equals ?

The formula for the distance D between a point (x, y) and a line $Ax + By + C = 0$ is, $D = \frac{|Ax_1 + By_1 + C|}{\sqrt{A^2 + B^2}}$.

Using this formula with $(x_1, y_1) = (k, 2k)$ and $A = 3$, $B = 4$, $C = (-)7$,

we get $D = \frac{|3(k) + 4(2k) - 7|}{\sqrt{3^2 + 4^2}} = 3$, or $(11k - 7) = \pm 15$.

Thus, $11k = 7 \pm 15$, or $k = 2$, $k = (-)0.73$.

ACT Math Reference – Formula and Notes

8. Arithmetic Series and Geometric Series:

- For Arithmetic series,

$a_n = a_1 + (n-1)d$, $d = a_n - a_{n-1}$, $s_n = \frac{a_1 + a_n}{2} \cdot n$, where d is the common difference.

(eg. 1) If the 10th term of an arithmetic sequence is 15 and the 20th term of the sequence is 100, what is the first term of the sequence?

Referring to the previous note on "Sequence and Series", where $a_n = a_1 + (n - 1)d$,
$a_{10} = a_1 + 9d = 15$, $a_{20} = a_1 + 19d = 100$
$\therefore a_{20} - a_{10} = 10d = 85 \quad \therefore d = 8.5$
$\therefore a_1 = 15 - 9d = 15 - 9(8.5) = (-)61.5$

- For Geometric Series;

$a_n = a_1 r^{n-1}$, $r = \frac{a_n}{a_{n-1}}$, $S_n = a_1 \cdot \frac{1 - r^n}{1 - r}$

(eg. 1) If the 3rd term of a geometric progression is $\sqrt[3]{a}$ and the 6th term is $\sqrt[3]{a^2}$, what is the 12th term of the progression?

Recall the Geometric series with the following formulas:

$a_n = a_1 r^{n-1}$, $r = \frac{a_n}{a_{n-1}}$, $S_n = a_1 \frac{1 - r^n}{1 - r}$

And also, the infinite series, $s = a_1 \frac{1}{1 - r}$, $|r| < 1$. Here, $a_3 = a^{\frac{1}{3}}$ and $a_6 = a^{\frac{2}{3}}$.

$\therefore \frac{a_6}{a_3} = \frac{a_1 r^5}{a_1 r^2} = r^3$, which is, $\frac{a^{\frac{2}{3}}}{a^{\frac{1}{3}}} = r^3$, or $a^{\frac{1}{3}} = r^3 \rightarrow r = a^{\frac{1}{9}}$. Therefore, $a_{12} = a_6 \cdot r^6 = a^{\frac{2}{3}} \cdot \left(a^{\frac{1}{9}}\right)^6 = a^{\frac{2}{3}} \cdot a^{\frac{2}{3}} = a^{\frac{4}{3}}$.

- For Infinite Series,

$s = a_1 \left(\frac{1}{1 - r}\right)$, $|r| < 1$

(eg. 1) Find the product of an infinite number of terms: $5^{\frac{1}{3}} \times 5^{\frac{2}{9}} \times 5^{\frac{4}{27}} \times 5^{\frac{8}{81}} \times \ldots$

$5^{\frac{1}{3}} \times 5^{\frac{2}{9}} \times 5^{\frac{4}{27}} \times 5^{\frac{8}{81}} \times \ldots = 5^{\frac{1}{3}(1 + \frac{2}{3} + \frac{4}{9} + \frac{8}{27} + \ldots)} = 5^{\frac{1}{3}\left[1 + \frac{2}{3} + \left(\frac{2}{3}\right)^2 + \left(\frac{2}{3}\right)^3 + \ldots\right]}$

Now, consider a geometric series: $1 + \frac{2}{3} + \left(\frac{2}{3}\right)^2 + \left(\frac{2}{3}\right)^3 + \ldots$ which is an infinite series with $a_1 = 1$, $r = \frac{2}{3}$

$\therefore$ The sum of the series, $S = \frac{1}{1 - \frac{2}{3}} = 3 \quad \therefore 5^{\frac{1}{3}(1 + \frac{2}{3} + \left(\frac{2}{3}\right)^2 + \left(\frac{2}{3}\right)^3 + \ldots)} = 5^{\frac{1}{3}(3)} = 5^1 = 5$

(eg. 2) What is the sum of the infinite geometric series $1 + \frac{1}{2} + \frac{1}{4} + \frac{1}{8} + \frac{1}{16} + \frac{1}{32} + \ldots$?

For a given Geometric Sequence; $a_1, a_2, a_3, \ldots, a_n, \ldots$, we get $a_n = a_1 r^{n-1}$ and $s_n = a_1 \frac{1 - r^n}{1 - r}$.

Furthermore, for an infinite series, we have the sum, $s = a_1 \frac{1}{1 - r}$, $|r| < 1$.

Here, since our $r = \frac{1}{2}$, and $|r| < 1$, we get $s = (1) \frac{1}{1 - \frac{1}{2}} = 2$

9. Logic:

There are 4 different logic cases; let's make a statement. If p, then q. Using mathematical symbol for this statement is, $p \rightarrow q$.

- Negation: If p, then NOT q, or, $p \rightarrow \sim q$, where "~" sign represents "NOT".
- Inverse: If NOT p, then NOT q, or $\sim p \rightarrow \sim q$.
- Converse: If q, then p, or $q \rightarrow p$.
- Contra Positive: If NOT q, then NOT p, or $\sim q \rightarrow \sim p$.

Of all these, the Contra Positive is the most important one to remember. Here, the Contra Positive is exactly the same statement as the original statement. Often, if we cannot prove the original statement, then we may use the logic of Contra Positive to prove the original statement.

(eg. 1) The inverse of $p \rightarrow \sim q$ is equivalent to?

The inverse of $p \rightarrow \sim q$ is equivalent to negation on both, or $\sim p \rightarrow \sim(\sim q)$, which is $\sim p \rightarrow q$, because $\sim(\sim q) = q$ itself. However, we don't have any answer choice for this. Therefore, making Contra Positive of $\sim p \rightarrow q$ by switching $\sim p$ and q, and negating both, we get $\sim q \rightarrow \sim(\sim p)$ or $\sim q \rightarrow p$.

(eg. 2) Which of the following statements is logically equivalent to: "If she exercises, she will pass the test."

(A) She passed the test; therefore, she exercised.
(B) She did not exercise; therefore, she will not pass the test.
(C) She did not pass the test; therefore she did not exercise.
(D) She will pass the test only if she exercises.
(E) None of the above.

In this problem, "logically equivalent" means to find "contra positive". Referring to our previous note on "logic", the contra positive of the statement $p \rightarrow q$ is $\sim q \rightarrow \sim p$, where "~" means negation.
∴ "If she exercise, she will pass the test." becomes "If she does not pass the test, then she did not exercise."

(eg. 3) An indirect proof of the statement "If $x < 0$, then $\sqrt{x}$ is not a real number" could begin with the assumption that

(A) $x = 0$ (B) $x > 0$ (C) $\sqrt{x}$ is real number (D) $\sqrt{x}$ is not a real number (E) x is nonnegative

"An indirect proof" really means "Contra Positive" in logic, which means, the statement "if P, then Q", implies the contra positive statement of "if not Q, then not P".
Applying this, we get, if $\sqrt{x}$ is a real number, then $x \geq 0$.

10. Binomial Expansion,

$(A + B)^n = \sum_{r=0}^{n} {}_nC_r A^{(n-r)} B^r$

(eg. 1) When $(a + b)^4$ is expanded, what is the coefficient of the third term?

Refer to Binomial Expansion,

$$(a + b)^n = \sum_{r=0}^{n} {}_nC_r a^{n-r} b^r$$

Using $n = 4$, $r = 2$, we get the coefficient of the third term, ${}_4C_2 = 6$.

(eg. 2) Which of the following could be a term in the expansion of $(a - b)^8$?

(A) $56a^5b^3$ (B) $-56a^5b^3$ (C) $56a^6b^2$ (D) $-56a^6b^2$ (E) $-56a^2b^6$

Referring to the previous note on "Binomial Expansion".
$(a + b)^n = \sum_{r=0}^{n} {}_nC_r a^{n-r} b^r$, $(a - b)^8 = \sum_{r=0}^{8} {}_8C_r a^{8-r} (-b)^r$.
Here, let $r = 3$, then we get ${}_8C_3 a^5(-b)^3 = -56a^5b^3$.

11. Trigonometry

A) 6 Basic Trig Functions;

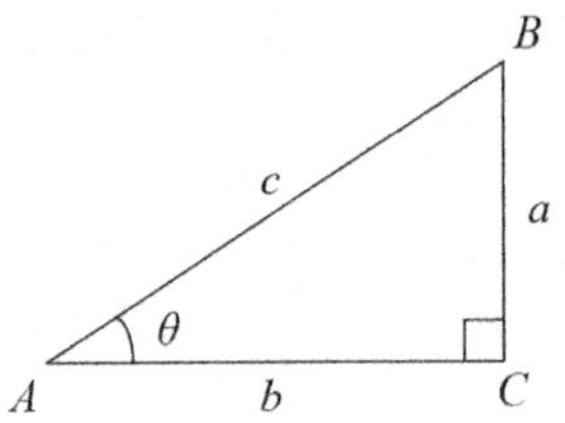

$\sin\theta = \dfrac{\text{opp}}{\text{hyp}} = \dfrac{a}{c}$, $\cos\theta = \dfrac{\text{adj}}{\text{hyp}} = \dfrac{b}{c}$, $\tan\theta = \dfrac{\text{opp}}{\text{adj}} = \dfrac{a}{b}$,

$\csc\theta = \dfrac{1}{\sin\theta}$, $\sec\theta = \dfrac{1}{\cos\theta}$, $\cot\theta = \dfrac{1}{\tan\theta}$

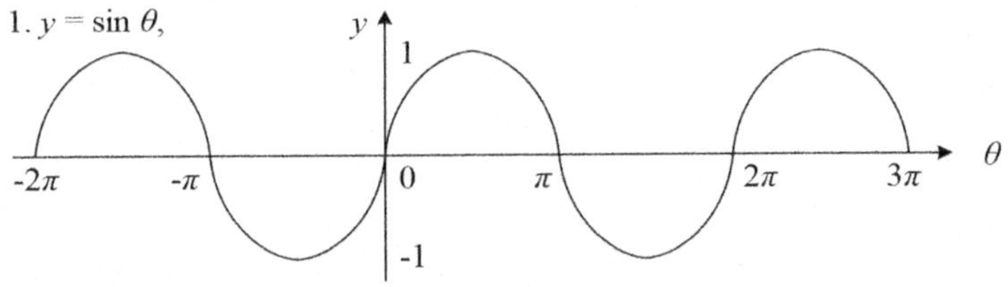

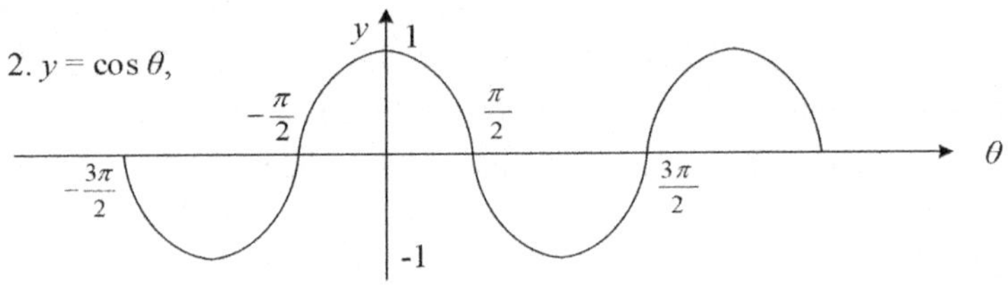

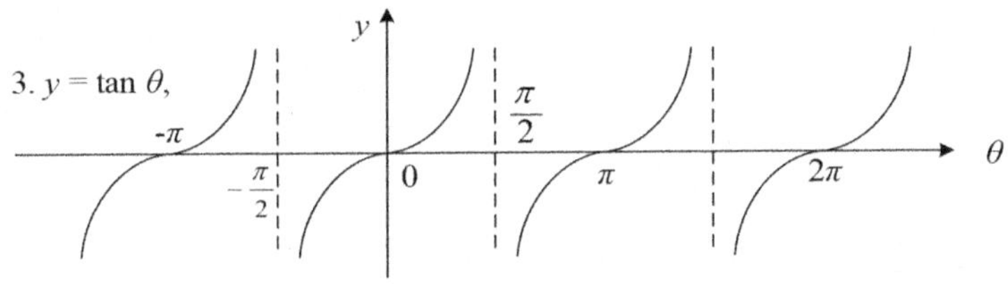

B) The property of the angle sum and difference

- $\sin(A \pm B) = \sin A\cos B \pm \cos A\sin B$
- $\cos(A \pm B) = \cos A\cos B \mp \sin A\sin B$
- $\tan(A \pm B) = \dfrac{\tan A \pm \tan B}{1 \mp \tan A\tan B}$
- $\sin 2A = 2\sin A\cos A$
- $\cos 2A = \cos^2 A - \sin^2 A = 1 - 2\sin^2 A = 2\cos^2 A - 1$
- $\sin^2 A + \cos^2 A = 1$
- $1 + \tan^2\theta = \sec^2\theta$

(eg. 1) cos (150° + x) – cos (150° - x) equals

(A) $\sqrt{2}\sin x$ (B) - $\sin x$ (C) $\sqrt{3}\cos x$ (D) $\sqrt{2}\cos x$ (E) $\sqrt{3}\sin x$

Refer to $\cos(A + B) = \cos A\cdot\cos B - \sin A\cdot\sin B$. Then $\cos(150° + x) - \cos(150° - x) =$

$(\cos 150°\cdot\cos x - \sin 150°\cdot\sin x) - (\cos 150°\cdot\cos x + \sin 150°\cdot\sin x) = (-)2\cdot\sin 150°\cdot\sin x = (-)2\cdot\frac{1}{2}\cdot\sin x = -\sin x$

But, for this problem, I strongly recommend to use a calculator by letting $x = 30°$ or some other angle.

Then $\cos(150° + 30°) - \cos(150° - 30°) = \cos 180° - \cos 120° = -1 - (-\frac{1}{2}) = -\frac{1}{2}$

Try answer choice (A) $\sqrt{2}\sin 30° = \sqrt{2}\frac{1}{2} = \frac{1}{\sqrt{2}}$ ∴ No!

(B) $-\sin 30° = -\frac{1}{2}$ ∴ Yes

(eg. 2) If $\sin x = \frac{3}{5}$ and $\frac{\pi}{2} \le x \le \pi$, then $\tan 2x$ =?

$\sin x = \frac{3}{5}$

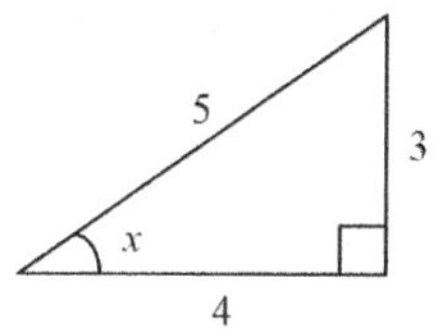

Now, from the triangle shown above, we get $\tan x = \frac{3}{4}$.

But our domain $\frac{\pi}{2} \le x \le \pi$ implies it is in the 2nd quadrant, where $\tan x = -\frac{3}{4}$.

Now, $\tan 2x = \frac{2\tan x}{1 - \tan^2 x} = \frac{2\left(-\frac{3}{4}\right)}{1 - \left(-\frac{3}{4}\right)^2} = \frac{-\frac{6}{4}}{\frac{7}{16}} = -\frac{6}{4} \times \frac{16}{7} = -\frac{24}{7}$

C) <u>Heron's formula to find the area, given 3 sides;</u>

Area $= \sqrt{s(s-a)(s-b)(s-c)}$, where $s = \frac{a+b+c}{2}$

(eg. 1) The area of a triangle with sides 4, 5, and 8 is

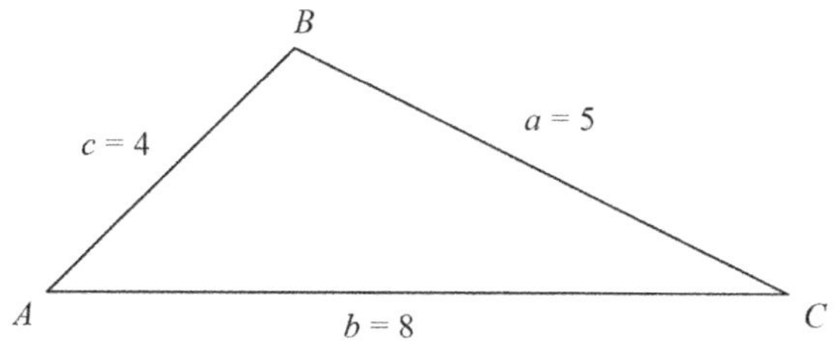

Here, we are given the lengths of 3 sides only, without any angle information.

Therefore, we use Heron's formula to find the area, that is $s = \frac{a+b+c}{2} = \frac{4+5+8}{2} = 8.5$

Area $= \sqrt{s(s-a)(s-b)(s-c)} = \sqrt{8.5(8.5-4)(8.5-5)(8.5-8)} = 8.18$

ACT Math Reference – Formula and Notes

D) Law of Sine and Law of Cosine

- The Law of Sine: $\frac{\sin A}{a} = \frac{\sin B}{b} = \frac{\sin C}{c}$
- The law of Cosine: $c^2 = a^2 + b^2 - 2ab \cdot \cos C$
- Area with 2 side, and angle θ: Area $= \frac{1}{2} \cdot bc \cdot \sin\theta$

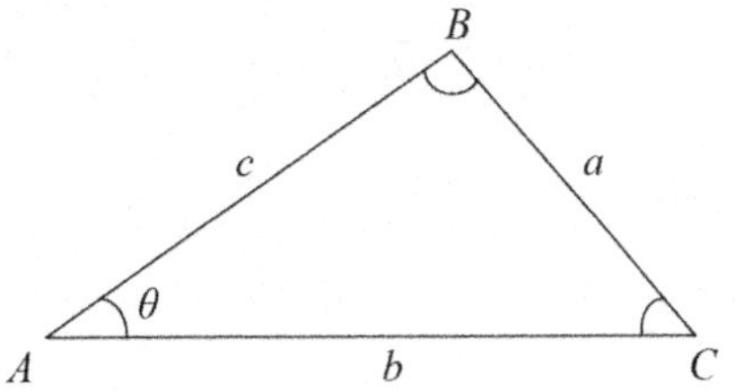

(eg. 1) What is the measure of the smallest angle in a right triangle with sides of lengths 5, 12 and 13?
Referring to the topics of Trig. Functions previously, we may use the Law of Cosine,
$c^2 = a^2 + b^2 - 2ab \cdot \cos C$. Here, the shortest side is 5.
$\therefore 5^2 = 12^2 + 13^2 - 2 \cdot 12 \cdot 13 \cdot \cos C$.
$\therefore C = \cos^{-1}\left(\frac{12^2 + 13^2 - 5^2}{2 \cdot 12 \cdot 13}\right) = 22.62°$

(eg. 2) What is the degree measure of the angle x of a triangle that has sides of length of 5 and 7, as shown in the figure (1)?

Using the Law of sine, we get $\frac{5}{\sin 30°} = \frac{7}{\sin x°}$

$\therefore \sin x = 7 \cdot \frac{\sin 30°}{5} = \frac{7}{10}$ $\quad \therefore x = \sin^{-1}\frac{7}{10} = 44.43°$

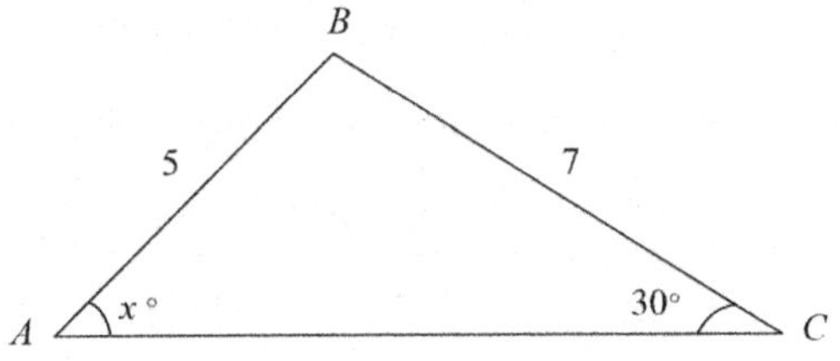

Figure (1)

(eg. 3) In $\triangle ABC$, $a = 2$, $b = 5$, $\angle C = 60°$.
What is the area of $\triangle ABC$?
Area of $\triangle ABC =$
$\frac{1}{2} \cdot a \cdot b \cdot \sin c° = \frac{1}{2}(2)(5) \cdot \sin 60° = \frac{5\sqrt{3}}{2}$.

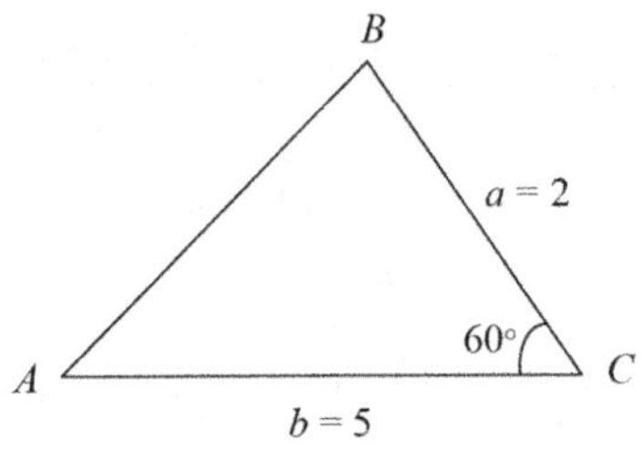

(eg. 4) In figure 37,
what is the approximate area of triangle ABC?

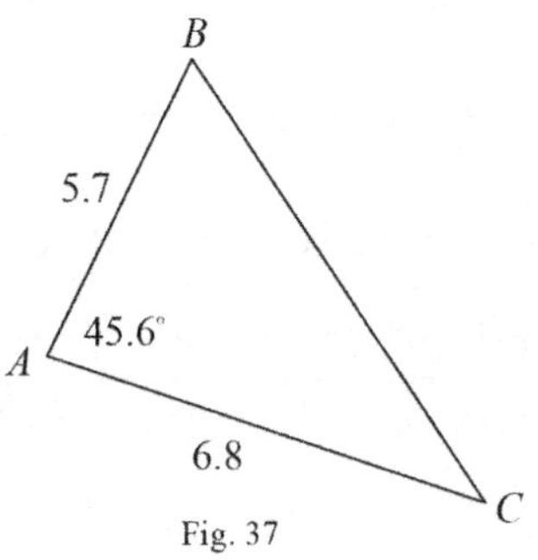

Fig. 37

The area of ΔABC is $\frac{1}{2}bh = \frac{1}{2}bc\sin\theta$, where $h = c\cdot\sin\theta$.
Since we have $b = 6.8$, $c = 5.7$ and $\theta = 45.6°$,
we have area $\Delta ABC = \frac{1}{2}\cdot(6.8)(5.7)\sin 45.6° = 13.85$.

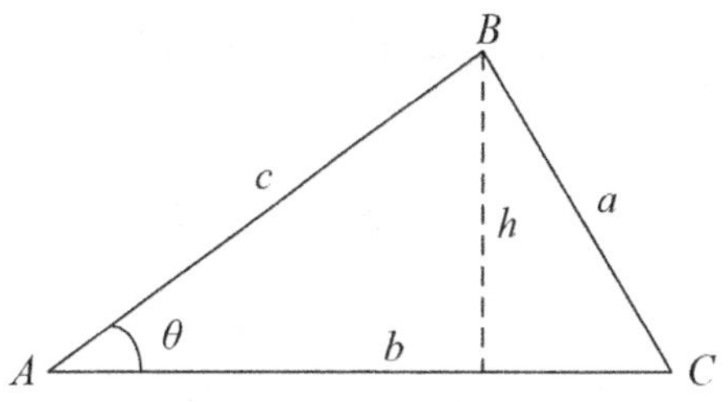

E) For a general form: $y = a\cos b(x - h) + k$,

we have, $|a|$ = Amplitude, b = # of frequency, $\frac{2\pi}{b}$ = period,
h = units of translation horizontally, k = units of shifting vertically.

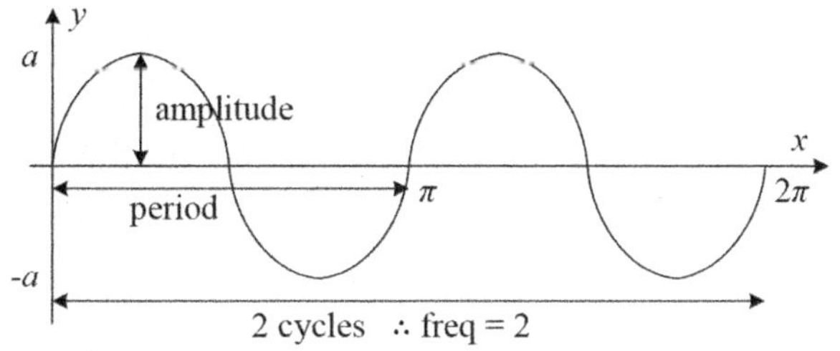

(eg. 1) What is the period of the graph of the function $y = \cos^2 x - \sin^2 x$?

Note that $\cos 2x = \cos^2 x - \sin^2 x$.

So, our equation becomes, $y = \cos^2 x - \sin^2 x = \cos 2x$, and thus, the period $p = \frac{2\pi}{2} = \pi$.

(eg. 2) What is the period of the curve whose equation is $y = 4\sin x\cdot\cos x$?

$y = 4\sin x\cdot\cos x = 2(2\sin x\cdot\cos x)$.
But $2\sin x\cdot\cos x = \sin 2x$.
$\therefore y = 2\cdot\sin 2x$.
$\therefore p = 2\frac{\pi}{2} = \pi = 180°$

F) Cofunctions

$\sin\theta = \cos(90° - \theta) = \cos(\frac{\pi}{2} - \theta)$, (ie) $\sin 30° = \cos 60°$

$\tan\theta = \cot(90° - \theta) = \cot(\frac{\pi}{2} - \theta)$, (ie) $\tan 15° = \cot 75°$

$\sec\theta = \csc(90° - \theta) = \csc(\frac{\pi}{2} - \theta)$, (ie) $\sec 45° = \csc 45°$

(eg. 1) If A is the degree measure of an acute angle and $\sin A = 0.25$, then $\sin A\cdot\cos(90° - A)$?
Recall the property of Cofunction. $\sin(90° - A) = \cos A$

(eg) $\sin 30° = \cos 60°$, $\tan 25° = \cot 65°$, $\sec 15° = \csc 75°$, ...
Therefore, $\sin A\cdot\cos(90° - A) = (\sin A)(\sin A) = (0.25)(0.25) = 0.0625$

(eg. 2) If $\tan A = \cot B$, which of the following must be true?
(A) $A = B$ (B) $A + B = 0$ (C) $A + B = 90°$ (D) $A + B = 180°$ (E) $A - B = 180°$
Since $\tan A = \cot B$, by the property (2), we must get $A + B = 90°$.

G) <u>Sign changes by Quadrants;</u> all seniors take calculus… by I-II-III-IV Qtr

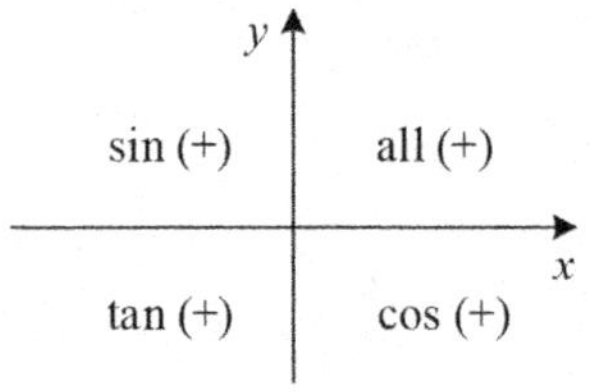

(eg. 1) If $(\csc x)(\cot x) < 0$, which of the following must be true?

I. $\tan x < 0$

II. $(\sin x)(\cos x) < 0$

III. x is in the second or third quadrant

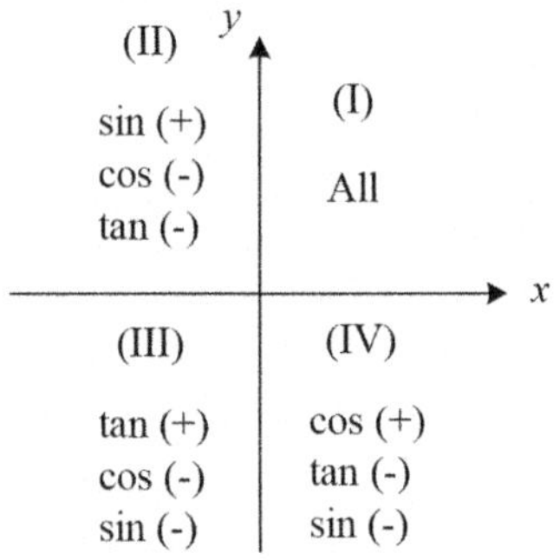

Remember that $\csc x = \frac{1}{\sin x}$ and $\cot x = \frac{1}{\tan x}$. Therefore, $(\csc x)(\cot x) = \frac{1}{\sin x} \times \frac{1}{\tan x} < 0$, which implies that the signs of $\sin x$ and $\tan x$ must be the opposite signs each other. Referring to the sign chart by quadrant, we have $\sin x$ and $\tan x$ have opposite signs each other in (II) and (III) quadrants.

(eg. 2)

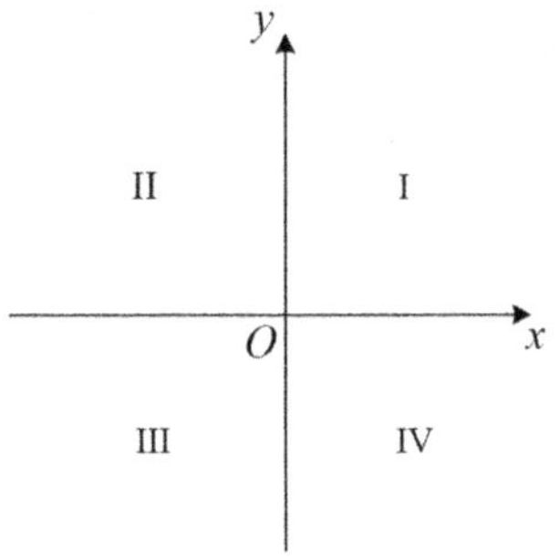

If $\sin\theta\cdot\cos\theta < 0$ and $\sin\theta\cdot\tan\theta > 0$, then θ must be in which quadrant in the figure above?

$\sin\theta\cdot\cos\theta < 0$ ∴ $\sin\theta$ and $\cos\theta$ must have opposite signs! ∴ II or IV

$\sin\theta\cdot\tan\theta > 0$ ∴ $\sin\theta$ and $\tan\theta$ must have the same signs. ∴ I or IV

Now, IV is the only quadrant that satisfy both!!

ACT Math Reference – Formula and Notes

12. Probability and Combination;

- The total possible number of arrangement on n objects is $n!$

(eg) (A, B, C) = 3!, because A-B-C, A-C-B, B-A-C, B-C-A, C-A-B, C-B-A, total of 6 = 3!
(A, B, C, D) = 4!
(A, B, C, D, E) = 5! ... etc.

- The total number of arrangement on n objects, but with the repetition of k and r objects is $\frac{n!}{k!r!}$

(eg) COFFEE $\rightarrow \frac{\text{6 letters}}{(\text{2 F's})(\text{2 E's})} = \frac{6!}{2!2!}$

Minimum $\rightarrow \frac{\text{7 letters}}{(\text{3 m's})(\text{2 i's})} = \frac{7!}{3!2!}$

- The total number of arrangement on n objects, but with k group with n_1 and n_2, ... members in them is $k!(n_1!)(n_2!)\ldots$

(eg) A, B, C, D, E into 2 grouping of (A, B), (C, D, E) will be
(2! grouping) × (2! on A, B) × (3! on C, D, E) = (2!)(2!)(3!)

- Permutation is, with a consideration of "Ordering", or ${}_nP_r = \frac{n!}{(n-r)!}$

(eg) Suppose I choose 3 best scored students out of 10 students in my Math IIc class, and then buy them lunch, steak for the best, hamburger for the second best, and tacos for the third best.
Then, this will be a permutation, because "Order" matters by having different lunches with their rank,
thus ${}_{10}P_3 = \frac{10!}{(10-3)!} = \frac{10!}{7!}$

- Combination is, without a consideration of "Ordering", or ${}_nC_r = \frac{n!}{(n-r)!r!}$

(eg) Suppose that I choose 3 best scored students out of 10 students in my Math IIc class, and then buy them lunch, but this time, all hamburgers. Then, in this case, "order" does not matter, because they all eat hamburgers anyway, regardless of their rank. Thus, it is Combination, ${}_{10}C_3 = \frac{10!}{(10-3)!3!} = \frac{10!}{7!3!}$.
Now, notice that ${}_{10}C_3$ and ${}_{10}P_3$ are exactly the same, except that ${}_{10}C_3$ is being divided by 3!,
while ${}_{10}P_3$ is not. That is because in ${}_{10}C_3$ case, those 3 students selected will have the same lunch,
hamburger – hamburger – hamburger, or H - H - H → repetition of 3, therefore, being divided by 3!.

ACT Math Reference – Formula and Notes

13. Conic Sections:

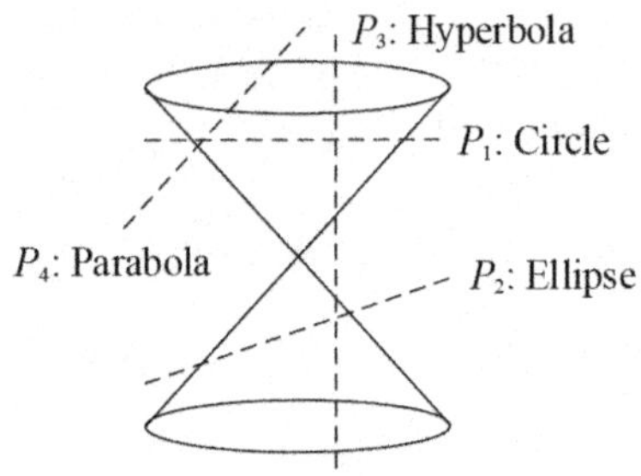

1) Circle: Let P_1 be the plane that cuts the cone in a horizontal direction. Then, it creates a circle, formed by its intersection.

Given $(x-h)^2+(y-k)^2=r^2$,
- center = (h, k)
- radius = r

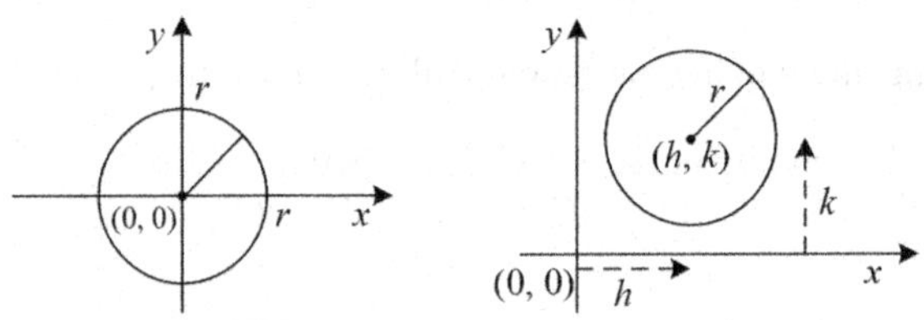

(eg. 1) Which ordered number pair represents the center of the circle $x^2+y^2+12x-8y=0$?
To get the center of this circle, we need to modify this eq. into a complete square form.
That is, $x^2+12x+y^2-8y=0$ becomes $(x+6)^2+(y-4)^2=52$. Therefore, we get the center $(-6, 4)$.

(eg. 2) The circumference of circle $x^2+y^2-6x-4y-36=0$ is
$x^2-6x+y^2-4y=36 \rightarrow$ making a complete square form, $(x-3)^2+(y-2)^2=36+9+4=49=7^2$
$\therefore$ center $c=(3, 2)$, radius $r=7$ $\therefore$ circumference $c=2\pi r=2\pi(7)=14\pi=43.98$

2) Ellipse: Let P_2 be the plane that cuts the cone in a tilted direction.

Then, it creates an elliptical figure, formed by intersection.

Given $\dfrac{(x-h)^2}{a^2}+\dfrac{(y-k)^2}{b^2}=1$
- center = (h, k)
- length of Major axis = $2a$, when $a > b$
- length of Minor axis = $2b$
- vertex = $(\pm a+h, o+k)$
- foci = $(\pm c+h, o+k)$, where $c^2=a^2-b^2$
- $d_1+d_2=K=2a$

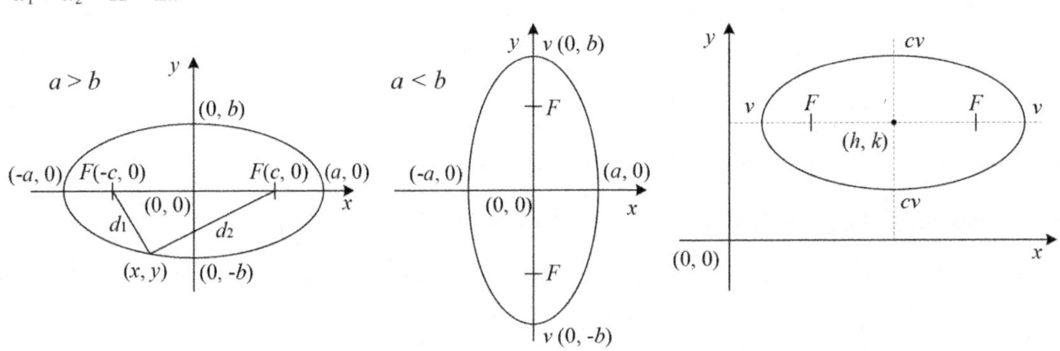

ACT Math Reference – Formula and Notes

(eg. 1) What is the length of the major axis of the ellipse whose equation is $10x^2 + 6y^2 = 15$?

$$10x^2 + 6y^2 = 15 \rightarrow \frac{10x^2}{15} + \frac{6y^2}{15} = 1$$

$$\rightarrow \frac{x^2}{\frac{15}{10}} + \frac{y^2}{\frac{15}{6}} = 1 \rightarrow \frac{x^2}{(\sqrt{1.5})^2} + \frac{y^2}{(\sqrt{2.5})^2} = 1$$

$\therefore a = \sqrt{1.5},\ b = \sqrt{2.5}$

$\therefore$ The length of major axis $= 2b = 2\sqrt{2.5} = 3.16$

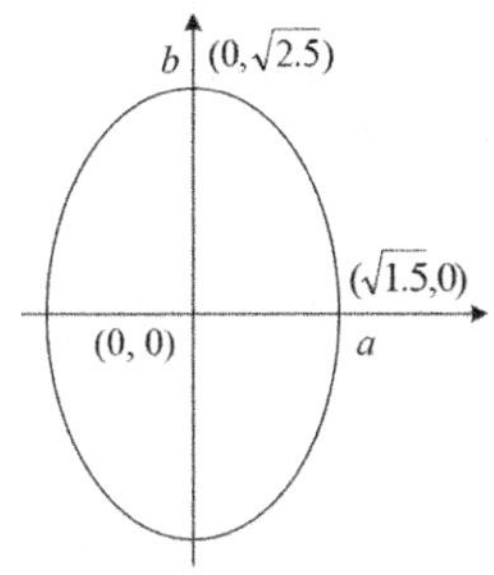

(eg. 2) If $3(x + 1)^2 + 10(y - 2)^2 = 48$ is graphed the sum of the distances from any point on the curve to the two foci is?

$$3(x + 1)^2 + 10(y - 2)^2 = 48 \rightarrow \frac{(x+1)^2}{16} + \frac{(y-2)^2}{4.8} = 1 \quad \therefore a = 4,\ b = \sqrt{4.8}$$

$\therefore$ The length of major axis $= 2a = 8$, which is equal to the sum of the distances between two fixed points to a point on the curve.

3) Hyperbola: Let P_3 be the plane that cuts the cone in a vertical direction.

Then, it creates a hyperbolic figure with 2 curves, formed by its intersection.

Given $\dfrac{(x-h)^2}{a^2} - \dfrac{(y-k)^2}{b^2} = 1$, or $\dfrac{(y-k)^2}{b^2} - \dfrac{(x-h)^2}{a^2} = 1$

- center = (h, k)
- vertex = $(\pm a + h, o + k)$, or $(0 + h, \pm b + k)$
- asymptotes: $y - k = \pm \frac{b}{a}(x - h)$, where the slope $\frac{b}{a}$ is $\frac{\text{rise}}{\text{run}}$.
- foci = $(\pm c + h, o + k)$ or $(0 + h, \pm c + k)$, where $c^2 = a^2 + b^2$
- $d_1 - d_2 = K = 2a$

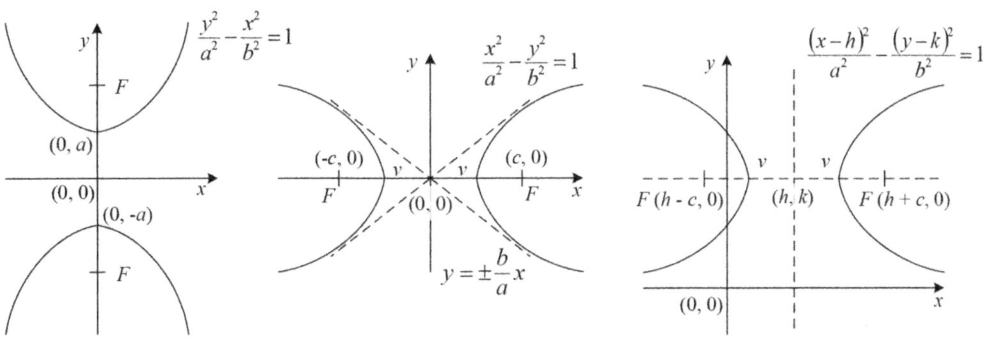

(eg. 1) When drawn on the same set of axes, the graphs of $x^2 - y^2 = 4$ and $4(x - 1)^2 + y^2 = 25$ have in common exactly?

(A) 0 points
(B) 1 point
(C) 2 points
(D) 3 points
(E) 4 points

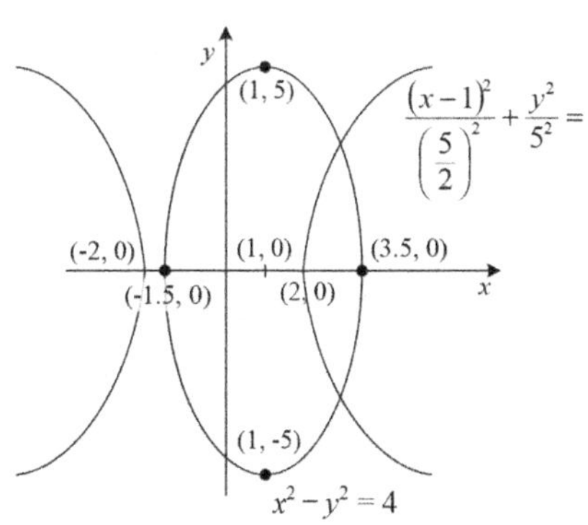

4) Parabola: Let P_4 be the plane that cuts the cone <u>parallel to slant edge direction</u>.

Then, it creates a parabolic figure with 1 curve, formed by its intersection.
Given $4p(y-k) = (x-h)^2$, or $4p(x-h) = (y-k)^2$,

- vertex = (h, k)
- foci = $(o + h, p + k)$ or $(p + h, o + k)$
- Directrix line: $y = -p + k$ or $x = -p + h$

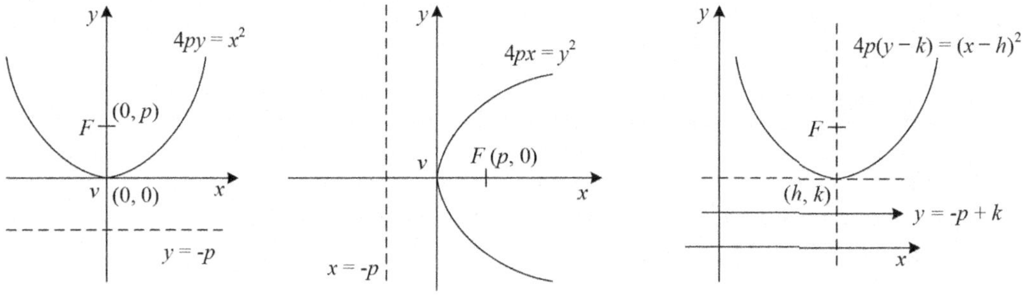

(eg. 1) The focus of a parabola is the point (2,0) and its directrix is the line $x = -2$. Write an equation of the parabola.
Referring to our previous note on "<u>Conic Sections</u>", the equation of parabola is $4py = x^2$ or $4px = y^2$.
Since directrix line is $x = (-)2$, we use $4px = y^2$, where the focus point is on x-axis with $F = (p, 0) = (2, 0)$.
$\therefore p = 2$ $\therefore 4(2)x = y^2$, which also have directrix $x = -p = -2$.

14. The ratios between a line and area and volume

Length:	1	→ 2	→ 3
Area:	1^2	→ 2^2	→ 3^2
Volume:	1^3	→ 2^3	→ 3^3

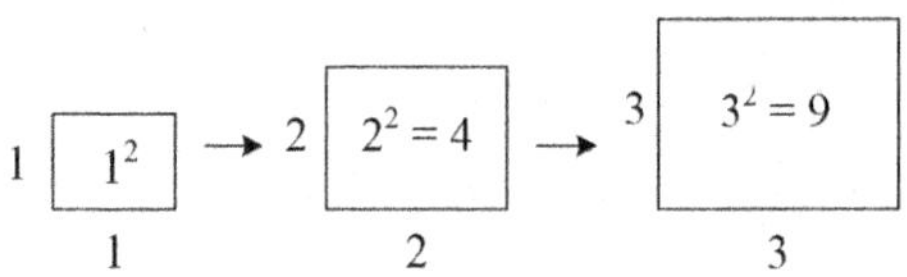

(eg. 1) One side of a given triangle is 6 inches. Inside the triangle a line segment is drawn parallel to this side, cutting off a triangle whose area is one-thirds that of the given triangle. Find the length of this segment in inches.

Since the ratio of the area ΔDBE is $1:\frac{1}{3}$, we get the ratio of the line segment $\overline{AC}:\overline{DE} = 1:\sqrt{\frac{1}{3}}$.

(Note)

Length:	1	→ 2	→ 3
Area:	1^2	→ 2^2	→ 3^2
Volume:	1^3	→ 2^3	→ 3^3

15. Matrix and Determinant;

• Let $A = \begin{pmatrix} a_{11} & a_{12} \\ a_{21} & a_{22} \end{pmatrix}$ $B = \begin{pmatrix} b_{11} & b_{12} \\ b_{21} & b_{22} \end{pmatrix}$, then, $A \pm B = \begin{pmatrix} a_{11} \pm b_{11} & a_{12} \pm b_{12} \\ a_{21} \pm b_{21} & a_{22} \pm b_{22} \end{pmatrix}$.

Then, $AB = \begin{pmatrix} a_{11} & a_{12} \\ a_{21} & a_{22} \end{pmatrix}\begin{pmatrix} b_{11} & b_{12} \\ b_{21} & b_{22} \end{pmatrix}$ = (rows of A) × (columns of B) = $\begin{pmatrix} a_{11}b_{11} + a_{12}b_{21} & a_{11}b_{12} + a_{12}b_{22} \\ a_{21}b_{11} + a_{22}b_{21} & a_{21}b_{12} + a_{22}b_{22} \end{pmatrix}$

• $[\, m \times n \,][\, n \times p \,] = [\, m \times p \,]$, The column of matrix A must be equal to the rows of matrix B!!

• Inverse of A matrix, $A^{-1} = \dfrac{1}{\text{Det } A}\begin{pmatrix} a_{22} & -a_{12} \\ -a_{21} & a_{11} \end{pmatrix}$

• Det $A = \begin{vmatrix} a_{11} & a_{12} \\ a_{21} & a_{22} \end{vmatrix} = a_{11} \cdot a_{22} - a_{12} \cdot a_{21}$

• If Det $A = 0$, then no inverse matrix of A exists!!!

(eg. 1) If matrix A has dimensions $m \times n$ and matrix B has dimensions $n \times m$, where m and n are distinct positive integers, which of the following statements must be true?

I The product BA does not exist.
II. The product AB exists and has dimensions $m \times m$.
III. The product AB exists and has dimensions $n \times n$.

The product of the matrix must match the dimension of columns of the first matrix equal to the rows of the second matrix.

$\therefore [m \times n][n \times p] = [m \times p]$

Here, (I) $BA = [n \times m][m \times n] = [n \times n]$ $\therefore$ BA exist $\therefore$ (I) is not true!

(II) $AB = [m \times n][n \times m] = [m \times m]$ $\therefore$ AB exist with $[m \times m]$

(III) Not true!

16. Variance and Standard Deviation;

• Given data, $x_1, x_2, x_3, \ldots, x_n$, we get Mean $= \bar{x} = \dfrac{x_1 + x_2 + x_3 + \ldots + x_n}{n} = \dfrac{\sum_{i=1}^{n} x_i}{n}$

$$\text{Var} = \sigma^2 = \frac{[(x_1 - \bar{x})^2 + (x_2 - \bar{x})^2 + \ldots + (x_n - \bar{x})^2]}{n} = \frac{\sum_{i=1}^{n} (x_i - \bar{x})^2}{n}$$

$$\text{Standard Deviation} = \sigma = \sqrt{\text{Variance}} = \sqrt{\frac{\sum_{i=1}^{n} (x_i - \bar{x})^2}{n}}$$

(Note) The Standard Deviation measures the average of how far the data points are away from the Mean.

(eg. 1) Of the following lists of numbers, which has the biggest standard deviation?

(A) 2, 5,8 (B) 3, 5, 9 (C) 4, 6, 8 (D) 1, 9, 18 (E) 2, 8, 9

In statistics, the standard deviation means "how far each point is away from the mean". But for this problem, without even using the above formula, we intuitively know that the answer choice (D) 1, 9, 18 has the biggest difference between them. Therefore, (D) has the biggest standard deviation.

ACT Math Reference – Formula and Notes

17. Solid Geometry;

- Volume Sphere: $V = \frac{4}{3}\pi r^3$
- Surface Area Sphere: $S = 4\pi r^2$
- Volume Cone: $V = \frac{1}{3}\pi r^2 h$
- Lateral Area of Cone: L.A. $= \frac{1}{2} \cdot c \cdot l$, where $c = 2\pi r$, $l = \sqrt{r^2 + h^2}$
- Volume Cylinder: $V = \pi r^2 h$

(eg. 1) If the radius of a sphere is doubled, the percent increase in surface area is?

$V = \frac{4}{3}\pi r^3$, $SA = 4\pi r^2$.

Now, r is doubled. $\therefore 2r$

$\therefore V' = \frac{4}{3}\pi(2r)^3$, $SA' = 4\pi(2r)^2$

Since $SA' = 16\pi r^2$ vs. $SA = 4\pi r^2$, it is 4:1 ratio.

$\therefore$ the percent difference is 300%.

(eg. 2) If surface area of a sphere has the same numerical value as its volume, what is the length of the radius of this sphere?

The formula for the surface area of sphere is $S = 4\pi r^2$, and the volume $V = \frac{4}{3}\pi r^3$.

$\therefore S = V \rightarrow 4\pi r^2 = \frac{4}{3}\pi r^3$, or $1 = \frac{1}{3}r$

$\therefore r = 3$

(eg. 3) The radius of a sphere is equal to the radius of the base of the cone. The height of a cone is equal to the radius of its base. The ratio of the volume of the *cone* to the volume of the *sphere* is?

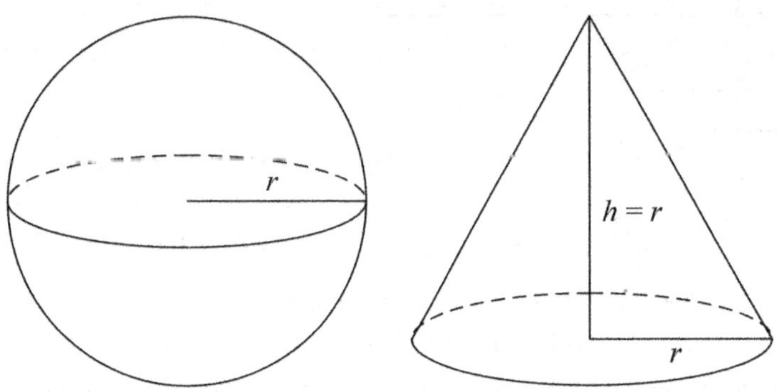

The volume of sphere is $V_1 = \frac{4}{3}\pi r^3$, and the volume of cone is $V_2 = \frac{1}{3}\pi r^2 h = \frac{1}{3}\pi r^2 (r) = \frac{1}{3}\pi r^3$

$$\therefore \frac{V_2}{V_1} = \frac{\frac{1}{3}\pi r^3}{\frac{4}{3}\pi r^3} = \frac{1}{4}$$

18. Arc length and area of pie sector;

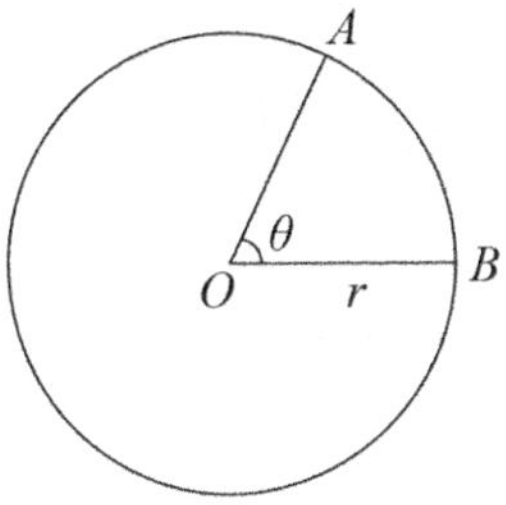

$\widehat{AB} = 2\pi r \times \frac{\theta}{360°} = 2\pi r \times \frac{\theta}{2\pi} = r \cdot \theta$ in radian mode.

Area $OAB = \pi r^2 \times \frac{\theta}{360°} = \pi r^2 \times \frac{\theta}{2\pi} = \frac{1}{2} r^2 \theta$ in radian mode.

(eg. 1) A sector of a circle, AQB, with a central angle of $\frac{\pi}{3}$ and a radius of 6 is bent to form a cone with vertex at Q. What is the volume of the cone that is formed?

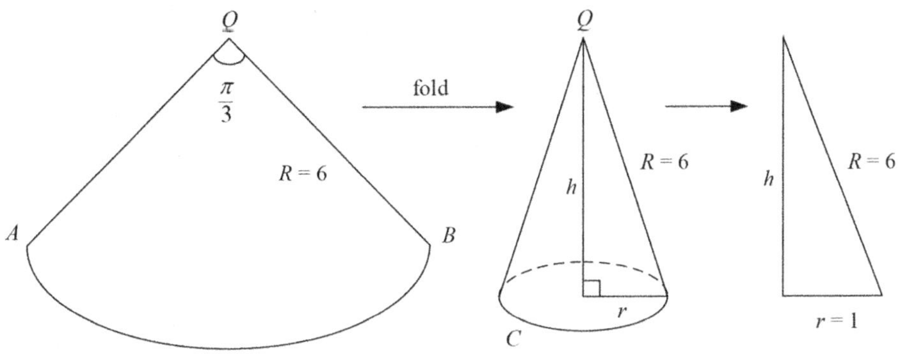

Figure (1) Figure (2) Figure (3)

To get the volume of the cone in the figure (2), we need to get the radius r and the height h, because $V = \frac{1}{3}\pi r^2 h$.
Notice that the arc length $\widehat{AB}$ becomes the circumference C of the cone.
But $\widehat{AB} = R \times \theta = 6 \times \frac{\pi}{3} = 2\pi$, and the formula for the circumference $C = 2\pi r$.
Since $\widehat{AB} = C$, or $2\pi = 2\pi r$, we get $r = 1$. Now, referring to figure (3), we get $h = \sqrt{6^2 - 1^2} = \sqrt{35}$.
Therefore, the volume of the cone is $V = \frac{1}{3}\pi r^2 h$ or $V = \frac{1}{3}\pi (1)^2 \cdot \sqrt{35} = 6.2$.

(eg. 2) A rod, pivoted at one end, rotates through $\frac{\pi}{4}$ radians. If the rod is 16 inches long, how many inches does the free end travel?

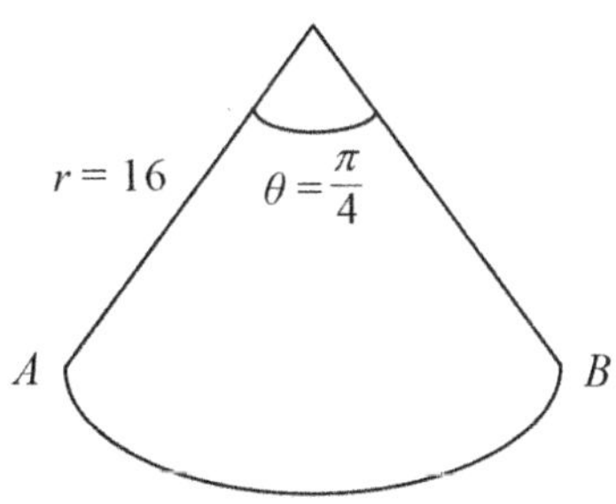

$\widehat{AB} = r \cdot \theta = 16 \frac{\pi}{4} = 4\pi$

Model Test No. 01

60 Questions / 60 Minutes

Directions: Solve each problem, choose the correct answer, and then fill in the corresponding oval on your answer document.

Do not linger over problems that take too much time. Solve as many as you can; then return to the others in the time you have left for this test.

You are permitted to use a calculator on this test. You many use your calculator for any problems you choose, but some of the problems may best be done without using a calculator.

Note: Unless otherwise stated, all of the following should be assumed.

1. Illustrative figures are NOT necessarily drawn to scale.
2. Geometric figures lie in a plane.
3. The word line indicates a straight line.
4. The word average indicates arithmetic mean.

Answer Sheet

Model Test No. 01

1	Ⓐ Ⓑ Ⓒ Ⓓ Ⓔ	16	Ⓕ Ⓖ Ⓗ Ⓙ Ⓚ	31	Ⓐ Ⓑ Ⓒ Ⓓ Ⓔ	46	Ⓕ Ⓖ Ⓗ Ⓙ Ⓚ
2	Ⓕ Ⓖ Ⓗ Ⓙ Ⓚ	17	Ⓐ Ⓑ Ⓒ Ⓓ Ⓔ	32	Ⓕ Ⓖ Ⓗ Ⓙ Ⓚ	47	Ⓐ Ⓑ Ⓒ Ⓓ Ⓔ
3	Ⓐ Ⓑ Ⓒ Ⓓ Ⓔ	18	Ⓕ Ⓖ Ⓗ Ⓙ Ⓚ	33	Ⓐ Ⓑ Ⓒ Ⓓ Ⓔ	48	Ⓕ Ⓖ Ⓗ Ⓙ Ⓚ
4	Ⓕ Ⓖ Ⓗ Ⓙ Ⓚ	19	Ⓐ Ⓑ Ⓒ Ⓓ Ⓔ	34	Ⓕ Ⓖ Ⓗ Ⓙ Ⓚ	49	Ⓐ Ⓑ Ⓒ Ⓓ Ⓔ
5	Ⓐ Ⓑ Ⓒ Ⓓ Ⓔ	20	Ⓕ Ⓖ Ⓗ Ⓙ Ⓚ	35	Ⓐ Ⓑ Ⓒ Ⓓ Ⓔ	50	Ⓕ Ⓖ Ⓗ Ⓙ Ⓚ
6	Ⓕ Ⓖ Ⓗ Ⓙ Ⓚ	21	Ⓐ Ⓑ Ⓒ Ⓓ Ⓔ	36	Ⓕ Ⓖ Ⓗ Ⓙ Ⓚ	51	Ⓐ Ⓑ Ⓒ Ⓓ Ⓔ
7	Ⓐ Ⓑ Ⓒ Ⓓ Ⓔ	22	Ⓕ Ⓖ Ⓗ Ⓙ Ⓚ	37	Ⓐ Ⓑ Ⓒ Ⓓ Ⓔ	52	Ⓕ Ⓖ Ⓗ Ⓙ Ⓚ
8	Ⓕ Ⓖ Ⓗ Ⓙ Ⓚ	23	Ⓐ Ⓑ Ⓒ Ⓓ Ⓔ	38	Ⓕ Ⓖ Ⓗ Ⓙ Ⓚ	53	Ⓐ Ⓑ Ⓒ Ⓓ Ⓔ
9	Ⓐ Ⓑ Ⓒ Ⓓ Ⓔ	24	Ⓕ Ⓖ Ⓗ Ⓙ Ⓚ	39	Ⓐ Ⓑ Ⓒ Ⓓ Ⓔ	54	Ⓕ Ⓖ Ⓗ Ⓙ Ⓚ
10	Ⓕ Ⓖ Ⓗ Ⓙ Ⓚ	25	Ⓐ Ⓑ Ⓒ Ⓓ Ⓔ	40	Ⓕ Ⓖ Ⓗ Ⓙ Ⓚ	55	Ⓐ Ⓑ Ⓒ Ⓓ Ⓔ
11	Ⓐ Ⓑ Ⓒ Ⓓ Ⓔ	26	Ⓕ Ⓖ Ⓗ Ⓙ Ⓚ	41	Ⓐ Ⓑ Ⓒ Ⓓ Ⓔ	56	Ⓕ Ⓖ Ⓗ Ⓙ Ⓚ
12	Ⓕ Ⓖ Ⓗ Ⓙ Ⓚ	27	Ⓐ Ⓑ Ⓒ Ⓓ Ⓔ	42	Ⓕ Ⓖ Ⓗ Ⓙ Ⓚ	57	Ⓐ Ⓑ Ⓒ Ⓓ Ⓔ
13	Ⓐ Ⓑ Ⓒ Ⓓ Ⓔ	28	Ⓕ Ⓖ Ⓗ Ⓙ Ⓚ	43	Ⓐ Ⓑ Ⓒ Ⓓ Ⓔ	58	Ⓕ Ⓖ Ⓗ Ⓙ Ⓚ
14	Ⓕ Ⓖ Ⓗ Ⓙ Ⓚ	29	Ⓐ Ⓑ Ⓒ Ⓓ Ⓔ	44	Ⓕ Ⓖ Ⓗ Ⓙ Ⓚ	59	Ⓐ Ⓑ Ⓒ Ⓓ Ⓔ
15	Ⓐ Ⓑ Ⓒ Ⓓ Ⓔ	30	Ⓕ Ⓖ Ⓗ Ⓙ Ⓚ	45	Ⓐ Ⓑ Ⓒ Ⓓ Ⓔ	60	Ⓕ Ⓖ Ⓗ Ⓙ Ⓚ

MATHEMATICS TEST

60 Minutes – 60 Questions

DIRECTIONS: Solve each problem, choose the correct answer, and then fill in the corresponding oval on your answer document.

Do not linger over problems that take too much time. Solve as many as you can; then return to the others in the time you have left for this test.

You are permitted to use a calculator on this test. You many use your calculator for any problems you choose, but some of the problems may best be done without using a calculator.

Note: Unless otherwise stated, all of the following should be assumed.

1. Illustrative figures are NOT necessarily drawn to scale.
2. Geometric figures lie in a plane.
3. The word line indicates a straight line.
4. The word average indicates arithmetic mean.

---

1. If $2x + 5y = 8$ and $5x - 7y = 13$, what is the value of $11x + 8y$?

A. 3
B. 8
C. 15
D. 18
E. 37

2. In the xy-plane, the points (3, 4), (-3, 2) and (p, 5) are collinear. What is the value of p?

F. 1
G. 3
H. 4
J. 6
K. 8

3. If x and y are positive integers and $4x + 3y = 13$, what is the value of x?

A. 1
B. 2
C. 3
D. 4
E. 5

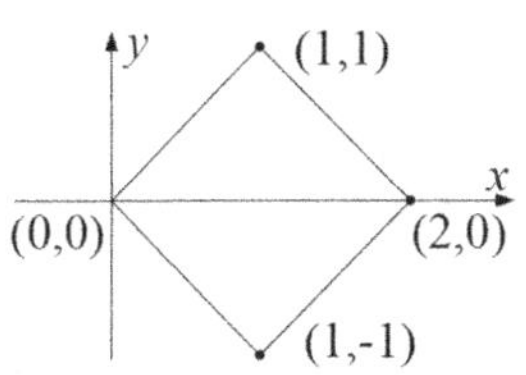

4. What is the area of the square shown above?

F. 2
G. $2\sqrt{2}$
H. 4
J. $\sqrt{2}$
K. $4\sqrt{2}$

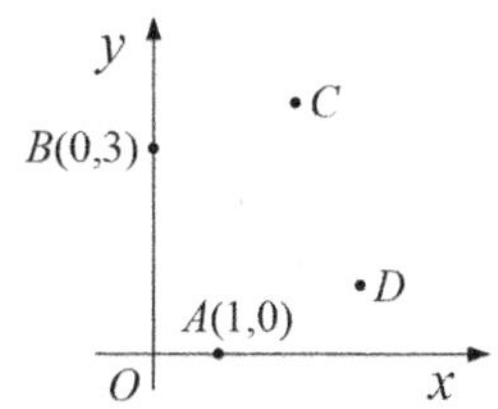

5. In the figure above, points A, B, C and D are vertices of a square. Which of the following are the coordinates of point C?

A. (3, 3)
B. (4, 3)
C. (2, 4)
D. (4, 2)
E. (3, 4)

6. If $2y \geq 5$ and $3y > 10$, which of the following could be the value of y?

F. 0
G. 1
H. 2
J. 3
K. 4

7. If the perimeter of a triangle ABC is p when $AB = AC = 2BC$, what is the length of $\overline{BC}$ in terms of p?

A. $\frac{1}{6}p$
B. $\frac{1}{5}p$
C. $\frac{1}{4}p$
D. $\frac{1}{3}p$
E. $\frac{1}{2}p$

GO ON TO THE NEXT PAGE.

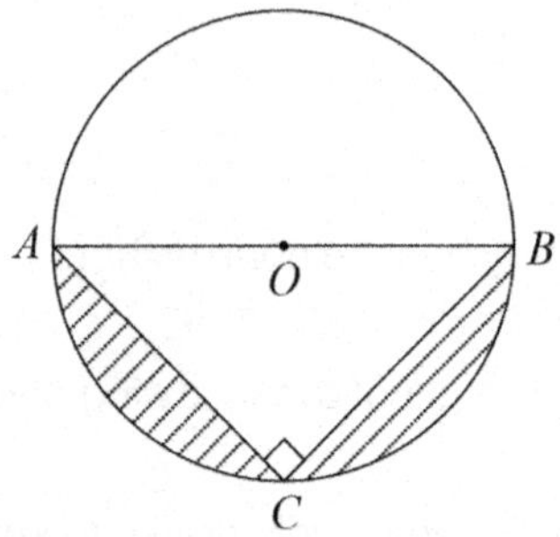

8. In the figure above, the circle with center O has a diameter AB of its length 8, and $AC = BC$. Which of the following is the area of the shaded part?

F. $8\pi - 16$
G. 8π
H. 16
J. $4\pi + 4$
K. 8

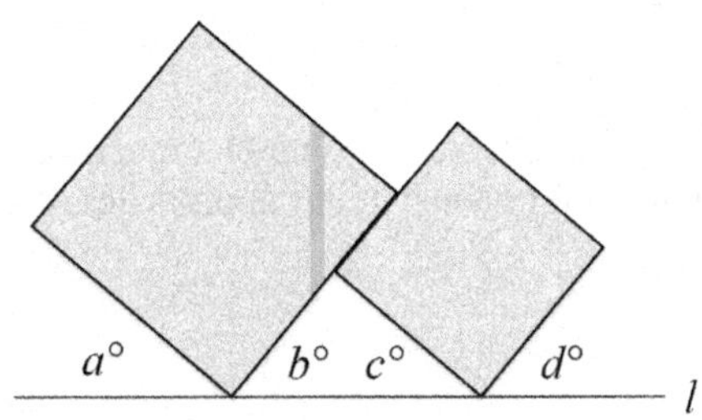

Note: Figure not drawn to scale.

9. In the figure above, the two shaded regions are squares. Which of the following must be true?

A. $a = b$
B. $a = d$
C. $b = c$
D. $b = d$
E. $c = d$

10. If 4 less than x is 1 more than 5, what is the value of x?

F. 3
G. 4
H. 7
J. 8
K. 10

11. If $p - 7 = q$, then p is 3 less than which of the following?

A. $q - 10$
B. $q - 4$
C. q
D. $q + 4$
E. $q + 10$

12. The length of each side of a rectangle will be tripled to create a second rectangle. The area of the second triangle will be how many times the area of the original rectangle?

F. 3
G. 6
H. 9
J. 12
K. 27

13. When 1,600 homing pigeons were released, only 400 reached the destination. What fraction of the pigeons that were released did not reach the destination?

A. $\frac{5}{6}$
B. $\frac{3}{4}$
C. $\frac{6}{7}$
D. $\frac{8}{9}$
E. $\frac{9}{10}$

14. The function f is defined by $f(x) = x^2 + bx - 28$, where b is a constant. In the xy-plane, the graph of $y = f(x)$ crosses the x-axis where $x = -4$. What is the value of b?

F. -7
G. -3
H. -2
J. 2
K. 5

15. If the function f is defined by $f(x) = 3x$, which of the following expressions is equal to $f\left(\frac{2}{3}x + 1\right) - 2$?

A. $2x + 1$
B. $2x + 2$
C. $3x$
D. $3x + 1$
E. $3x + 2$

16. If $(y - 2x)(2y - x) = k$, which of the following is always equal to $(x - 2y)(3y - 6x)$?

F. $2k$
G. $3k$
H. $4k$
J. $-3k$
K. $-2k$

GO ON TO THE NEXT PAGE.

17. If x and y are negative numbers, then the inequality $x\sqrt{2} < y\sqrt{3}$ must be equivalent to which of the following?

A. $x < \frac{3}{2}y$
B. $x > \frac{3}{2}y^2$
C. $|x| > \frac{3}{2}|y|$
D. $x^2 < \frac{3}{2}y^2$
E. $x^2 > \frac{3}{2}y^2$

18. One number is 6 less than twice another number. If the sum of the two numbers is 30, what is the greater of the two numbers?

F. 10
G. 12
H. 15
J. 18
K. 31

19. Positive integers x, y, and z satisfy the equations $x^y = 243$ and $y^z = 625$. What is the value of $x + z$?

A. 5
B. 7
C. 9
D. 11
E. 13

$|a + b| = |a| + |b|$

20. Which of the following pairs (a, b) shows that the equation above is <u>not</u> always true?

F. (0, 0)
G. (2, 2)
H. (-2, -2)
J. (-4, 0)
K. (4, -4)

21. White, black, yellow or red marbles are placed in a single line so that the pattern of white, black, yellow, red, white, black, yellow, red repeats throughout. If the first marble in the line is white, which of the following marbles is yellow?

A. 117th
B. 118th
C. 119th
D. 120th
E. 121st

```
  DVD
-  DV
-----
   CD
```

22. In the correctly worked subtraction problem above, each letter represents a different digit. What is the value of DVD + CD?

F. 91
G. 101
H. 111
J. 192
K. 202

23. A right cone has a base of circumference 6π. If the volume of the cone is 48π, what is the height?

A. 4
B. 8
C. 12
D. 16
E. 32

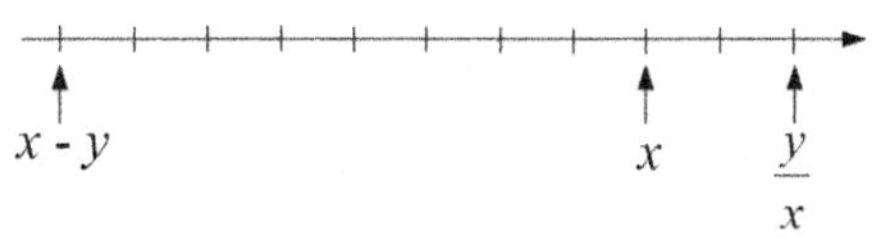

24. On the number line above, the tick marks are equally spaced. Which of the following represents the value of x?

F. -2
G. -1
H. 0
J. 2
K. 3

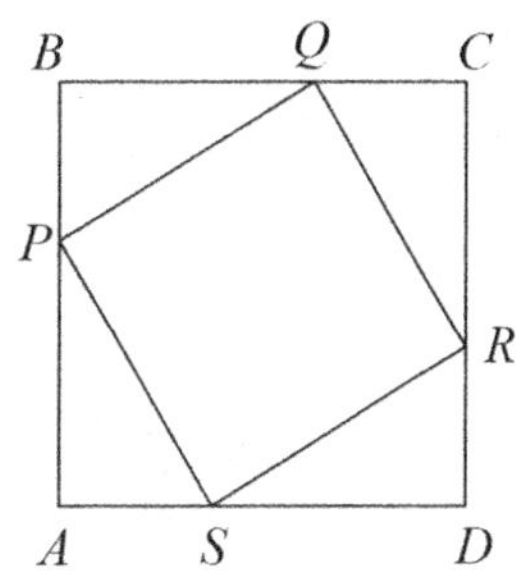

25. In the figure above, the vertices of square $PQRS$ lie on square $ABCD$. If square $ABCD$ has side of length 12 and $AS = \frac{1}{2}SD$, what is the perimeter of square $PQRS$?

A. $4\sqrt{5}$
B. $8\sqrt{5}$
C. 80
D. $16\sqrt{5}$
E. 160

GO ON TO THE NEXT PAGE.

26. What is the surface area of a cube if its volume is 216 cubic inches?

F. $27\sqrt{6}$ sq in
G. $36\sqrt{6}$ sq in
H. 64 sq in
J. 125 sq in
K. 216 sq in

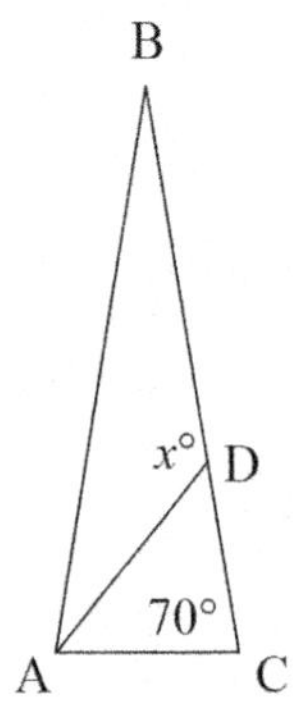

Note: Figure not drawn to scale.

27. In the figure above, if $AB = BC$ and $AD = BD$, then $x =$

A. 40
B. 50
C. 60
D. 80
E. 100

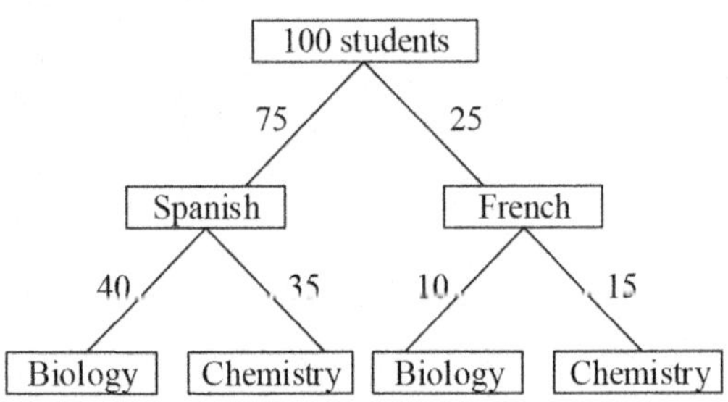

28. The chart above shows the number of students who are studying French, the number of students who are studying Spanish, and the number of students who are studying Biology or chemistry. Of the students who are studying chemistry, what fraction is studying Spanish?

F. $\frac{2}{5}$

G. $\frac{1}{2}$

H. $\frac{3}{5}$

J. $\frac{7}{10}$

K. $\frac{4}{5}$

29. A rectangular has a perimeter of 60 inches and a width of 18 inches. What is the area of the rectangular?

A. 212
B. 216
C. 224
D. 232
E. 248

30. There are 18 songs recorded on a compact disk. Half of the songs are 3 minutes long, one third of them are 5 minutes long and the rest are 2 minutes long. What is the average (arithmetic mean) number of minutes per song?

F. 3
G. 3.3
H. 3.5
J. 4
K. 4.2

31. A number x can be expressed as 75 less than 125 percent of itself. What is the value of x?

A. 250
B. 270
C. 275
D. 300
E. 315

32. A telephone call between Irvine and Inglewood costs 50 cents for the first minute and 20 cents for each additional minute. Amanda made a 10-minute call from Irvine to Inglewood. Her brother Jerry also made a call from Irvine to Inglewood, but his call cost three times as much as Amanda's. How long was Jerry's call, in minutes? (Assume that Jerry's call was a whole number of minutes.)

F. 27
G. 29
H. 30
J. 32
K. 33

33. The numerals of 0, 2, 3, 5 and 7 will be used to make 4-digit numbers. If no numerals can be used more than once, how many different 4-digit numbers can be made?

A. 48
B. 56
C. 64
D. 72
E. 96

GO ON TO THE NEXT PAGE.

34. Steve and Tom collect stamps. If Steve had 11 more stamps, he would have 2 times as many stamps as Tom has. Steve has 135 stamps. How many stamps do the two boys have altogether?

F. 208
G. 235
H. 246
J. 259
K. 281

35. The 2010 minimum wage was 40 percent more than the 1990 minimum wage. If the 2010 minimum wage was \$8.75 per hour, what was the 1990 minimum wage in dollars per hour?

A. 6.25
B. 6.50
C. 6.75
D. 7.00
E. 7.25

36. What is the average (arithmetic mean) of all the integers from -49 to 51, inclusive?

F. -0.5
G. 0
H. 0.5
J. 1
K. 1.5

37. If $x^2 - y^2 = 117$ and $x - y = 13$, what is the value of x?

A. 8
B. 10
C. 11
D. 13
E. 16

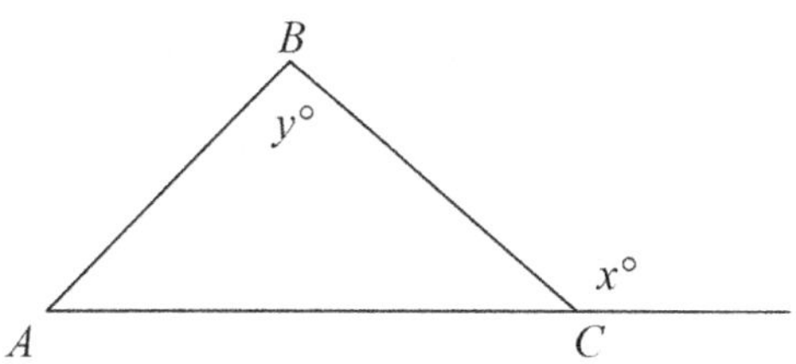

Note: Figure not drawn to scale.

38. In the figure above, $AB = BC$ and $x = 150°$. What is the value of y?

F. 100°
G. 110°
H. 120°
J. 130°
K. 140°

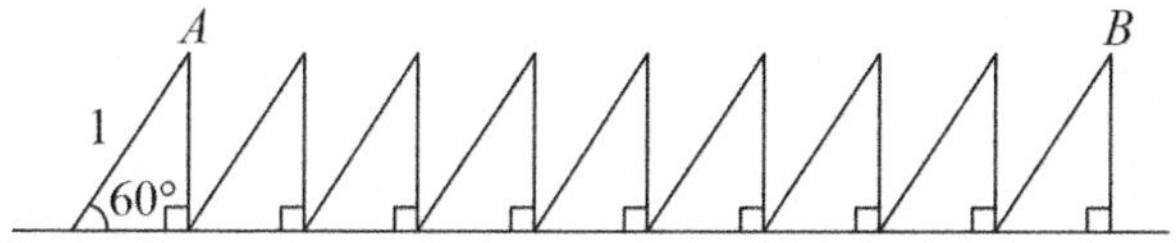

39. In the figure above, the bases of nine identical right triangles lie along a straight line. How far apart are points A and B?

A. 4
B. $4\sqrt{2}$
C. $4\sqrt{3}$
D. 5
E. It cannot be determined from the information given.

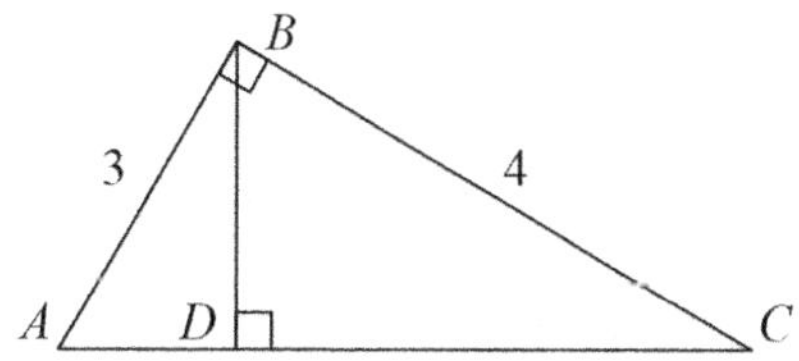

40. In the figure above, $\overline{AB} \perp \overline{BC}$, $\overline{BD} \perp \overline{AC}$, $AB = 3$ and $BC = 4$. What is the perimeter of $\triangle ABD$?

F. 7.2
G. 7.4
H. 8.6
J. 9.6
K. 12

41. At his normal swimming speed, Eugene swims a distance of d meters in 4 minutes. If he swims only at $\frac{2}{3}$ of his normal swimming speed, how many minutes will it take him to swim a distance of $\frac{d}{2}$ meters?

A. $\frac{4}{3}$
B. 2
C. $\frac{8}{3}$
D. 3
E. $\frac{10}{3}$

42. Eight points in a plane are arranged so that no 3 of them are collinear. What is the total number of straight lines connecting two points?

F. 24
G. 28
H. 12
J. 16
K. 18

GO ON TO THE NEXT PAGE.

$a < b < c$
$d < e < f$
$e < a$
$f < c$

43. If each letter in the inequalities above represents a positive integer, then, of the following fractions, which is the smallest?

A. $\frac{a}{e}$
B. $\frac{d}{8}$
C. $\frac{d}{b}$
D. $\frac{c}{f}$
E. $\frac{d}{c}$

44. If a nonzero number n is divided by its reciprocal, what is the result?

F. 1
G. n
H. $2n$
J. n^2
K. $2n^2$

45. If 20 percent of 60 percent of a positive number is equal to w percent of the same number, what is the value of w?

A. 1,200
B. 120
C. 12
D. 1.2
E. 0.12

46. During a sale, a customer can buy one shirt for x dollars. The second shirt and the rest of the shirts that the customer buys is discounted to $\frac{x}{2}$ dollars. For example, the cost of 5 shirts is $3x$ dollars. Which of the following represents the customer's cost, in dollars, for $2n$ shirts bought during this sale?

F. $\left(\frac{2n - 3}{2}\right)x$
G. $\left(\frac{2n - 1}{2}\right)x$
H. $(2n)x$
J. $(2n - 1)x$
K. $\left(\frac{2n + 1}{2}\right)x$

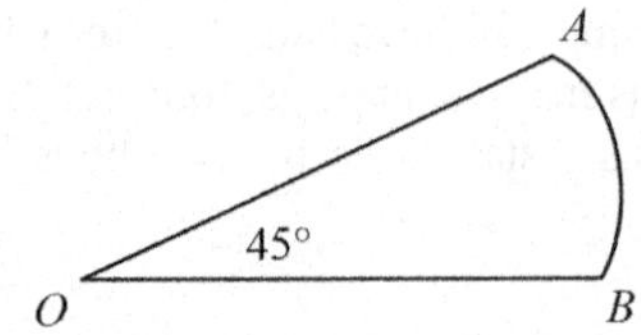

47. In the figure above, $\overset{\frown}{AB}$ is the arc of a circle with center O. If the length of arc $\overset{\frown}{AB}$ is 3π, what is the area of region OAB?

A. 45π
B. 36π
C. 27π
D. 18π
E. 9π

48. A water tank measuring 3 yards by 4 yards by 2 yards is filled to capacity by a pipe in $\frac{1}{3}$ hour. How many hours will it take to fill a water tank measuring 6 yards by 8 yards by 3 yards to capacity, if it is filled by the same pipe with water flowing at the same rate?

F. 1
G. 2
H. 4
J. 8
K. 13.5

49. Five distinct lines in a plane are such that no two lines are parallel and no three lines intersect at the same point. What is the total number of points of intersection of these five lines?

A. Four
B. Five
C. Six
D. Eight
E. Ten

50. Jim is working t hours each day and spends $\frac{t}{5}$ hours each day for a marketing project. In terms of t, how many days will it take Jim to complete the project that require a total of 20 hours?

F. $\frac{t}{100}$
G. $\frac{t}{20}$
H. $\frac{20}{t}$
J. $\frac{100}{t}$
K. $20t$

GO ON TO THE NEXT PAGE.

51. If $t > 0$, which of the following is equivalent to 25 percent of t?

A. $t - 0.75$
B. $t - 0.25$
C. $\frac{t}{25}$
D. $\frac{t}{75}$
E. $\frac{t}{4}$

52. A positive integer is called a palindrome if it reads the same forward as it does backward. For example, 959 and 8228 are palindromes, whereas 1332 is not. Neither the first nor the last digit of palindrome can be 0. How many palindromes are there between 100 and 1,500?

F. 94
G. 95
H. 104
J. 105
K. 1,400

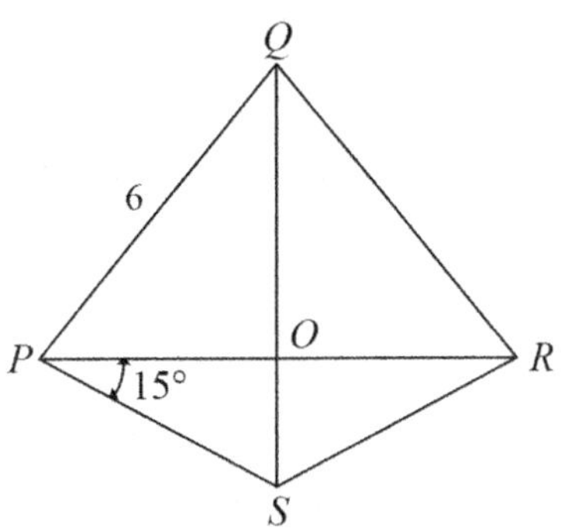

Note: Figure not drawn to scale.

53. In the quadrilateral above, if $PQ = SQ = RQ$, $PS = SR$ and $PQ = 6$, then $PR =$

A. $2\sqrt{3}$
B. $3\sqrt{2}$
C. 4
D. 5
E. 6

54. Stack of 20 boxes of equal height is 4 feet high. What is the height, in inches, of each box?

F. 4.8
G. 2.4
H. 3.6
J. 1.2
K. 2

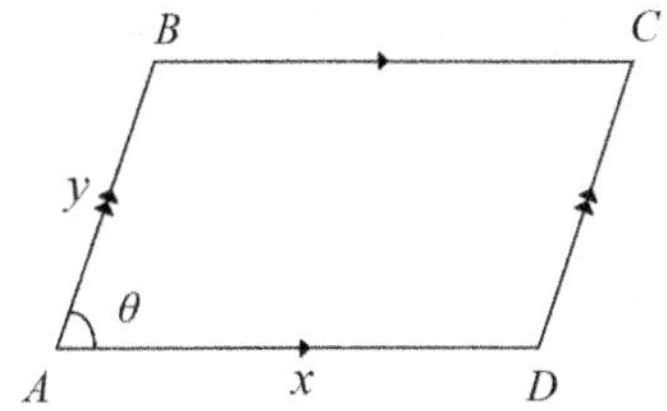

55. What is the area of the parallelogram $ABCD$ in above figure, in terms of x, y, and θ?

A. $\frac{1}{2}xy$
B. xy
C. $xy \cos \theta$
D. $xy \sin \theta$
E. $xy \tan \theta$

56. If the graph of $f(x) = x^2 + 1$ is translated 3 units to the right and then 4 units down, what would be the value of the transformed graph at the point $x = 4$?

F. 1.5
G. 5.1
H. -7.2
J. -2.0
K. -2.5

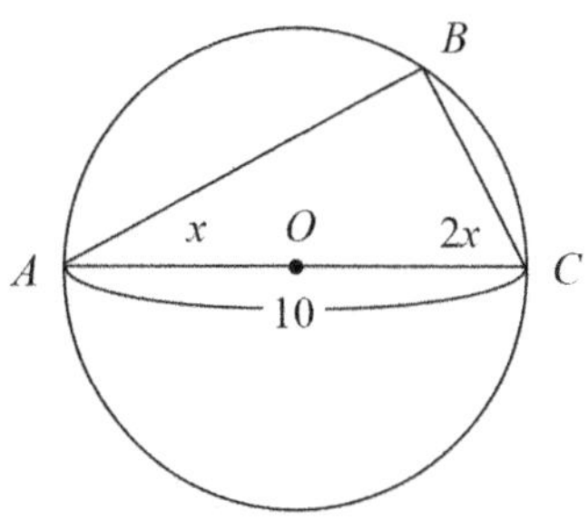

57. In above figure, what is the area of ΔABC?

A. 21.65
B. 20.21
C. 20.19
D. 20.17
E. 18.01

58. If $x_0 = 1$ and $x_{n+1} = (x_n - 1)^2$, what is the value of x_3?

F. -3
G. -2
H. -1
J. 0
K. 1

GO ON TO THE NEXT PAGE.

59. If $p(x) = ax^3 + bx^2 + cx + d$ such that $p(0) = 1$, $p(1) = 3$, and $p(-1) = 5$, then the value of b is

A. 3
B. 2
C. 1
D. 0
E. Not enough information to tell

60. If $f(x) = \frac{1}{1+x^2}$ and $g(x) = \sqrt{x-4}$, then what is the value of $f(g(13))$?

F. $\frac{1}{4}$
G. $\frac{1}{8}$
H. $\frac{2}{3}$
J. $\frac{1}{5}$
K. $\frac{1}{10}$

END OF TEST 1
STOP! DO NOT TURN THE PAGE UNTIL TOLD TO DO SO.
DO NOT RETURN TO THE PREVIOUS TEST.

Answer Key
Model Test No. 01

1	E	16	J	31	D	46	K
2	J	17	D	32	K	47	D
3	A	18	J	33	E	48	G
4	F	19	B	34	F	49	E
5	E	20	K	35	A	50	J
6	K	21	C	36	J	51	E
7	B	22	J	37	C	52	G
8	F	23	D	38	H	53	E
9	D	24	J	39	A	54	G
10	K	25	D	40	F	55	D
11	E	26	K	41	D	56	J
12	H	27	E	42	G	57	A
13	B	28	J	43	E	58	J
14	G	29	B	44	J	59	A
15	A	30	H	45	C	60	K

How to Score the ACT Mathematics Test

When you take an actual ACT Mathematics Test, your answer sheet will be "read" by a scanning machine that will record your responses to each question. Then a computer will compare your answers with the correct answers and produce your raw score. You get one point for each correct answer. For each wrong answer, there is no penalty. Therefore, you must guess and answer all the questions to get a higher score. This raw score is converted to a scaled score that is reported to you and to the colleges you specify.

Finding Your Scale Test Score

STEP 1: Count the number of right answers.
Enter the total here: ______________________

STEP 2: The number you obtained in Step 1 is your raw score.
Look up the conversion table for ACT-Math, and find your score.

Scaled Score Conversion Table
ACT Mathematics Test

Scale Score	Mathematics
36	60
35	58-59
34	56-57
33	55
32	54
31	52-53
30	50-51
29	48-49
28	46-47
27	43-45
26	41-42
25	39-40
24	37-38
23	35-36
22	33-34
21	31-32
20	30
19	27-29
18	25-26
17	22-24
16	18-21
15	15-17
14	12-14
13	09-11
12	08
11	06-07
10	05
09	04
08	-
07	03
06	02
05	-
04	-
03	01
02	-
01	00

1. If $2x + 5y = 8$ and $5x - 7y = 13$, what is the value of $11x + 8y$?

A. 3 **B.** 8 **C.** 15 **D.** 18 **E.** 37

(1) $2x + 5y = 8$
(2) $5x - 7y = 13$
Now, to get $11x + 8y$, let's multiply eq. (1) by 3, and then add to eq. (2), resulting 37.

Ans. E

2. In the xy-plane, the points (3, 4), (-3, 2) and (p, 5) are collinear. What is the value of p?

F. 1
G. 3
H. 4
J. 6
K. 8

Here, one way to find the value of p is by using the fact that the slope of the points between (3, 4) and (-3, 2), and (3, 4) and (p, 5) must be the same, because they are collinear. Therefore, using the formula slope $m = \frac{(y_2 - y_1)}{(x_2 - x_1)}$, we get $\frac{(4 - 2)}{(3 - (-3))} = \frac{(5 - 4)}{(p - 3)}$.
By cross multiplication, we get $2(p - 3) = 6$.
$\therefore p - 3 = 3 \quad \therefore p = 6$

Ans. J

3. If x and y are positive integers and $4x + 3y = 13$, what is the value of x?

A. 1 **B.** 2 **C.** 3 **D.** 4 **E.** 5

Note that x and y are integers. There are 2 unknowns and only one equation is provided. Therefore, there are infinitely many solutions. But since x and y are integers, you may try $x = 1, 2, 3$ ... So, let's try $x = 1$, then you get $y = 3$ only it works. But if you try $x = 2, 3, 4$ or 5, you may not get integer value for y. The only choice is $x = 1$.

Ans. A

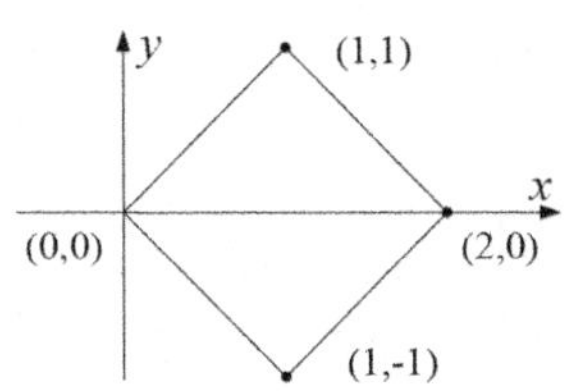

4. What is the area of the square shown above?

F. 2 **G.** $2\sqrt{2}$ **H.** 4 **J.** $\sqrt{2}$ **K.** $4\sqrt{2}$

One way to solve this problem guidely is, to notice that it is a square with its diagonal length of 2.

Therefore, the area is $\frac{1}{2} \times (2)(2) = 2$.

Ans. F

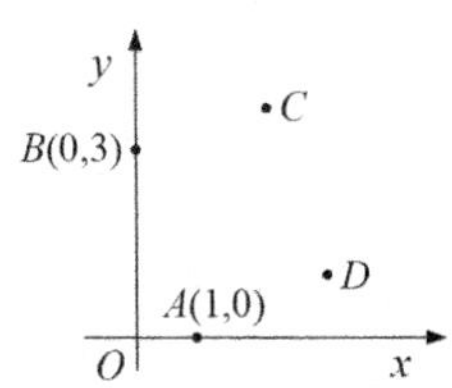

5. In the figure above, points A, B, C and D are vertices of a square. Which of the following are the coordinates of point C?

A. (3, 3) **B.** (4, 3) **C.** (2, 4) **D.** (4, 2) **E.** (3, 4)

Here, again it is a square. Therefore, the length of each side is equal. Using the formula, $D = \sqrt{(x_2 - x_1)^2 + (y_2 - y_1)^2}$ with coordinates of C being (x_2, y_2), we have $\overline{OB} = \overline{BC}$, or $\sqrt{(1 - 0)^2 + (0 - 3)^2} = \sqrt{(x_2 - 0)^2 + (y_2 - 3)^2}$.
That is, $10 = x_2^2 + (y_2 - 3)^2$.
The only correct answer is **E**, (3, 4).

Ans. E

6. If $2y \geq 5$ and $3y > 10$, which of the following could be the value of y?

F. 0
G. 1
H. 2
J. 3
K. 4

By solving $2y \geq 5$ and $3y > 10$, we get $y \geq \frac{5}{2}$ and $y > \frac{10}{3}$.

The common interval that satisfy both inequalities is $y > \frac{10}{3}$.

Therefore, the only choice is **K**, 4.

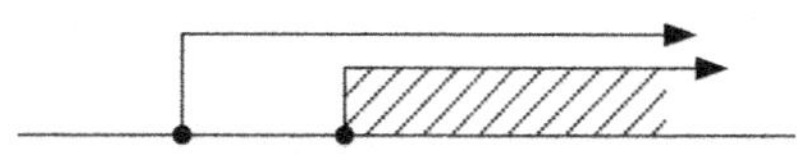

Ans. K

7. If the perimeter of a triangle ABC is p when $AB = AC = 2BC$, what is the length of $\overline{BC}$ in terms of p?

A. $\frac{1}{6}p$ **B.** $\frac{1}{5}p$ **C.** $\frac{1}{4}p$ **D.** $\frac{1}{3}p$ **E.** $\frac{1}{2}p$

Let the length of $\overline{BC}$ equals x, then $\overline{AB} = \overline{AC} = 2x$.
Therefore, total length of all 3 sides equals $5x$, and $p = 5x$.
Thus, $\overline{BC} = x = \frac{p}{5}$.

Ans. B

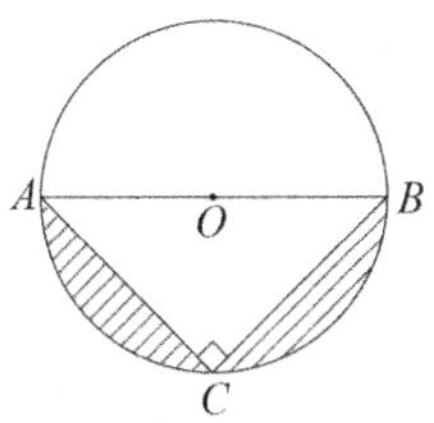

8. In the figure above, the circle with center O has a diameter AB of its length 8, and $AC = BC$. Which of the following is the area of the shaded part?

F. $8\pi - 16$ **G.** 8π **H.** 16 **J.** $4\pi + 4$ **K.** 8

The area of the shaded part is the difference of the semi-circle and the triangle ABC. The radius of circle is 4. The area of semi-circle is, $\frac{1}{2} \times (\pi r^2) = \frac{1}{2} \times \pi \times 42 = 8\pi$.

Also, the area of the triangle is, $A = \frac{1}{2} \times \text{Base} \times \text{Height}$.

Here, the base is the diameter 8, and the height is the same length as its radius. Therefore, $A = \frac{1}{2} \times (8) \times (4) = 16$.

$\therefore$ The shaded area is, $8\pi - 16$.

Ans. F

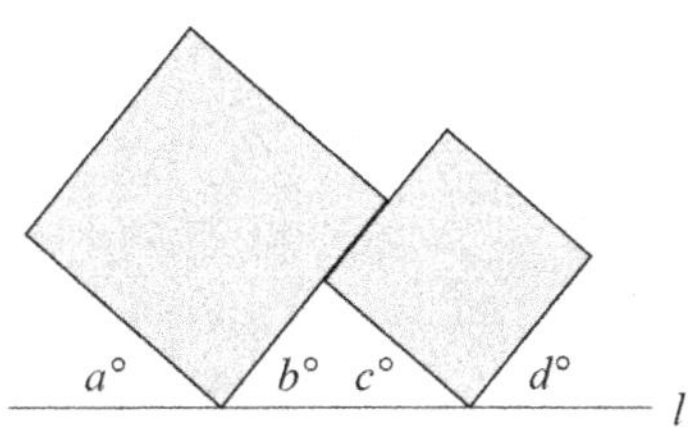

9. In the figure above, the two shaded regions are squares. Which of the following must be true?

A. $a = b$ **B.** $a = d$ **C.** $b = c$ **D.** $b = d$ **E.** $c = d$

Since the shaded regions are squares with all interior angles of 90°, and thus their sides become parallel to each other. We have angle $b°$ have the same measure as angle d °, because they are on parallel sides.

Ans. D

10. If 4 less than x is 1 more than 5, what is the value of x?

F. 3 **G.** 4 **H.** 7 **J.** 8 **K.** 10

The equation becomes, $x - 4 = 5 + 1$. Therefore, $x = 10$.

Ans. K

11. If $p - 7 = q$, then p is 3 less than which of the following?

A. $q - 10$ **B.** $q - 4$ **C.** q **D.** $q + 4$ **E.** $q + 10$

$p - 7 = q$. Therefore, $p = q + 7$.
Since p is 3 less than the value that we are looking for, Say S.
Then $p = S - 3$ or $S = p + 3$.
Since $p = q + 7$, we get $S = (q + 7) + 3 = q + 10$.

Ans. E

12. The length of each side of a rectangle will be tripled to create a second rectangle. The area of the second triangle will be how many times the area of the original rectangle?

F. 3 **G.** 6 **H.** 9 **J.** 12 **K.** 27

Referring to the following diagram,

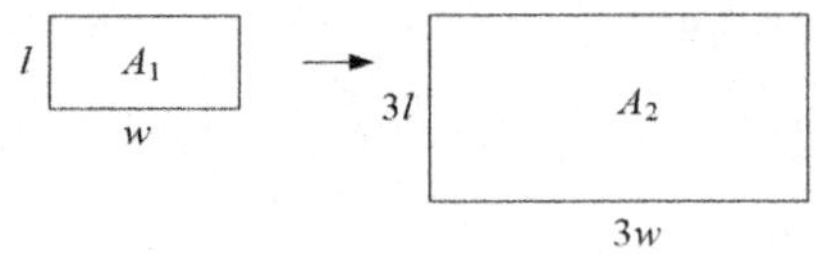

Then, $A_2 = 3l \times 3w = 9 \times (l \times w) = 9A$.

Ans. H

13. When 1,600 homing pigeons were released, only 400 reached the destination. What fraction of the pigeons that were released did <u>not</u> reach the destination?

A. $\frac{5}{6}$ **B.** $\frac{3}{4}$ **C.** $\frac{6}{7}$ **D.** $\frac{8}{9}$ **E.** $\frac{9}{10}$

Out of 1,600 pigeons, 400 reached the destination, while 1,200 did not. Therefore, the fraction of <u>NOT reached</u> will be 1,200 out of 1,600. $\left(\frac{1200}{1600}\right) = \frac{3}{4}$.

Ans. B

14. The function f is defined by $f(x) = x^2 + bx - 28$, where b is a constant. In the xy-plane, the graph of $y = f(x)$ crosses the x-axis where $x = -4$. What is the value of b?

F. -7 **G.** -3 **H.** -2 **J.** 2 **K.** 5

Since $x = (-)4$ is the x-intercept for $y = 0$, we solve $f(-4) = (-4)^2 + b(-4) - 28 = 0$, which results in $b = (-)3$.

Ans. G

15. If the function f is defined by $f(x) = 3x$, which of the following expressions is equal to $f\left(\frac{2}{3}x + 1\right) - 2$?

A. $2x + 1$
B. $2x + 2$
C. $3x$
D. $3x + 1$
E. $3x + 2$

$f(x)$ is defined by 3 times the input. Here, our input is $\left(\frac{2}{3}x + 1\right)$, therefore, $f\left(\frac{2}{3}x + 1\right) - 2 = 3 \times \left(\frac{2}{3}x + 1\right) - 2$
$= 2x + 3 - 2 = 2x + 1$.

Ans. A

16. If $(y - 2x)(2y - x) = k$, which of the following is always equal to $(x - 2y)(3y - 6x)$?

F. $2k$ **G.** $3k$ **H.** $4k$ **J.** $-3k$ **K.** $-2k$

$(x - 2y)(3y - 6x)$
$= (-)(2y - x) \times 3(y - 2x)$
$= (-)3 \times (2y - x)((y - 2x)$
$= (-)3k$

Ans. J

17. If x and y are negative numbers, then the inequality $x\sqrt{2} > y\sqrt{3}$ must be equivalent to which of the following?

A. $x < \frac{3}{2}y$ **B.** $x > \frac{3}{2}y^2$

C. $|x| > \frac{3}{2}|y|$ **D.** $x^2 < \frac{3}{2}y^2$ **E.** $x^2 > \frac{3}{2}y^2$

Since $x\sqrt{2} > y\sqrt{3}$, we get $x > \frac{\sqrt{3}}{\sqrt{2}}y$.
Now since x and y are both negative numbers, squaring both sides will result the change of the direction of the inequality.
Therefore, $(x\sqrt{2})^2 < (y\sqrt{3})^2$, or $2x^2 < 3y^2$, which results $x^2 < \frac{3}{2}y^2$.

Ans. D

18. One number is 6 less than twice another number. If the sum of the two numbers is 30, what is the greater of the two numbers?

F. 10 **G.** 12 H. 15 **J.** 18 **K.** 31

Applying mathematical translation, let x = one number and y = another number. Then, $x = (-)6 + 2y$. Now, the sum of the two numbers is, $x + y = 30$. Solving by substitute $x = -6 + 2y$ into the eq. $x + y = 30$, we get $(-6 + 2y) + y = 30$, which results in $3y = 36$. Therefore, $y = 12$, $x = 18$. The greater number is $x = 18$.

Ans. J

19. Positive integers x, y, and z satisfy the equations $x^y = 243$ and $y^z = 625$. What is the value of $x + z$?

A. 5 **B.** 7 **C.** 9 **D.** 11 **E.** 13

Remember x, y and z are all positive integers. By applying prime factorization of 243 and 625, we get $243 = 3^5$ and $625 = 5^4$. Therefore, setting $x^y = 3^5 = 243$and $y^z = 5^4 = 625$, we get $x = 3$, $y = 5$ and $z = 4$, which results $x + z = 3 + 4 = 7$.

Ans. B

$$|a + b| = |a| + |b|$$

20. Which of the following pairs (a, b) shows that the equation above is <u>not</u> always true?

F. (0, 0) **G.** (2, 2) **H.** (-2, -2) **J.** (-4, 0) **K.** (4, -4)

Let's plug in all of the answer choices **F**, **G**, **H**, **J** and **K**. The only choice that does not work is answer choice **K**, since $| 4 + (-4) | = 0$, whereas $| 4 | + | -4 | = 8$.

Ans. K

21. White, black, yellow or red marbles are placed in a single line so that the pattern of white, black, yellow, red, white, black, yellow, red repeats throughout. If the first marble in the line is white, which of the following marbles is yellow?

A. 117th **B.** 118th **C.** 119th **D.** 120th **E.** 121st

The pattern of repetition is, w – b – y – r. Therefore, every 4 has a cycle of repetition. Now, we want to get yellow, which is the <u>third one</u> of the cycle. Since it has a cycle of 4, we may interpret this as a multiple of 4's, and the third one as a remainder when divided by a cycle of 4. Answer choice 119th gives the remainder of 3, when it is divided by 4. (i.e.) $119 \div 4 = 29$ plus remainder of 3.

Ans. C

$$\begin{array}{r} \text{DVD} \\ - \quad \text{DV} \\ \hline \text{CD} \end{array}$$

22. In the correctly worked subtraction problem above, each letter represents a different digit. What is the value of DVD + CD?

F. 91 **G.** 101 **H.** 111 **J.** 192 **K.** 202

In this problem, please note that 3 digit number minus 2 digit number results in 2 digit number. Therefore, it only happens when DVD is a number in the range of 101, 111, 121…191, 198, thus D = 1. The only choice that result CD with its one digit D = 1, is when DVD = 101, because 111 is not a different digit and 121, 131… don't work, because 121 – 12 = 109, 131 – 13 = 118, and they are not two digit results!! Therefore, DVD – DV = 101 – 10 = 91, which is CD. Our answer is DVD + CD = 101 + 91 = 192.

Ans. J

23. A right cone has a base of circumference 6π. If the volume of the cone is 48π, what is the height?

A. 4 **B.** 8 **C.** 12 **D.** 16 **E.** 32

Since the base of cone has a circumference of 6π, we get $2\pi r = 6$. From this eq., we get $r = 3$. Now, the volume of the cone is $\frac{1}{3}\pi r^2 h$ and the equation becomes, $V = \frac{1}{3}\pi r^2 h = \frac{1}{3}\pi(3)^2 h = 48\pi$, which results in the height $h = 16$.

Ans. D

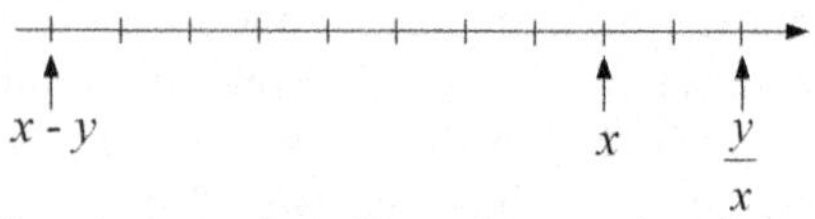

24. On the number line above, the tick marks are equally spaced. Which of the following represents the value of x?

F. -2 **G.** -1 **H.** 0 **J.** 2 **K.** 3

Since the tick marks of $(x - y)$ and x has their difference of 8 spaces, we get $x - (x - y) = 8$, resulting $y = 8$. Also, the difference between the tick marks of x and $\frac{y}{x}$ is 2 spaces, we get $\frac{y}{x} - x = 2$. But substituting $y = 8$ into the equation, we get $\frac{8}{x} - x = 2$, by multiplying x on both sides of the equation, we get $8 - x^2 = 2x$. That is, $x^2 + 2x = 8$ or $(x + 4)(x - 2) = 0$. x is either -4 or 2. Here, the only choice is Ans. **J**.

Ans. J

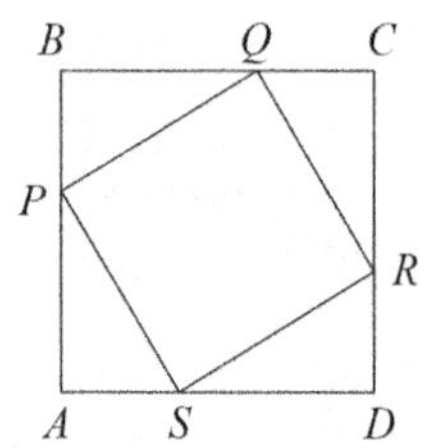

25. In the figure above, the vertices of square $PQRS$ lie on square $ABCD$. If square $ABCD$ has side of length 12 and $AS = \frac{1}{2}SD$, what is the perimeter of square $PQRS$?

A. $4\sqrt{5}$ **B.** $8\sqrt{5}$ **C.** 80 **D.** $16\sqrt{5}$ **E.** 160

In the figure, let $\overline{AS} = x$, then $\overline{SD} = 2x$.
But since the length of each side of the square $ABCD$ is equal to 12, we get $x + 2x = 12$, or $x = 4$.
Also, notice that the triangles ΔPAS, ΔSDR, ΔRCQ and ΔQBP are all congruent right triangles. Consider the triangle ΔPAS, which has side lengths of $\overline{AS} = 4$ and $\overline{PA} = 8$.
This results in the length of its hypotenuse
$\overline{PS} = \sqrt{(4^2 + 8^2)} = \sqrt{80} = 4\sqrt{5}$.
Therefore, the perimeter of square $PQRS$ is $4 \times 4\sqrt{5} = 16\sqrt{5}$.

Ans. D

26. What is the surface area of a cube if its volume is 216 cubic inches?

F. $27\sqrt{6}$ sq in **G.** $36\sqrt{6}$ sq in **H.** 64 sq in
J. 125 sq in **K.** 216 sq in

Let each side of cube is x.
Then the volume of the cube is $V = x^3 = 216$, or $x = 6$.
The surface area of cube is, $SA = 6x^2 + 6(6)^2 = 216$.

Ans. K

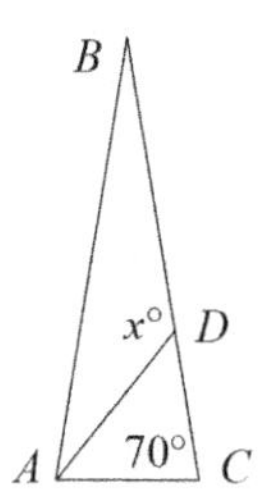

27. In the figure above, if $AB = BC$ and $AD = BD$, then $x =$

A. 40 **B.** 50 **C.** 60 **D.** 80 **E.** 100

Since ΔABC is an isosceles triangle, the base angles are the same. $\angle A = \angle C = 70°$, thus, $\angle B = 40°$.
Also, ΔABD is another isosceles triangle, which makes $\angle B = \angle DAB = 40°$. Therefore, $\angle ADB = x = 100°$.

Ans. E

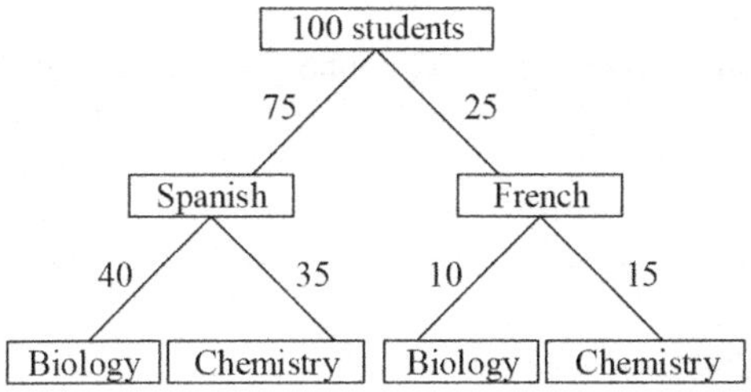

28. The chart above shows the number of students who are studying French, the number of students who are studying Spanish, and the number of students who are studying Biology or chemistry. Of the students who are studying chemistry, what fraction is studying Spanish?

F. $\frac{2}{5}$ **G.** $\frac{1}{2}$ **H.** $\frac{3}{5}$ **J.** $\frac{7}{10}$ **K.** $\frac{4}{5}$

Referring the diagram, the total number of students, studying chemistry is, $35 + 15 = 50$. Among them, there are 35 students coming from Spanish.
Therefore, the fraction of students studying chemistry out of the total chemistry students is, $\frac{35}{50}$ or $\frac{7}{10}$.

Ans. J

29. A rectangular has a perimeter of 60 inches and a width of 18 inches. What is the area of the rectangular?

A. 212 **B.** 216 **C.** 224 **D.** 232 **E.** 248

Let L = the length of the rectangle, and W = the length of width.
Then, the perimeter $p = 2L + 2W = 60$, where $W = 18$.
Therefore, $2L + 2(18) = 60$, or $L = 12$.
The area $A = L \text{ x } W = 12 \text{ x } 18 = 216$.

Ans. B

30. There are 18 songs recorded on a compact disk. Half of the songs are 3 minutes long, one third of them are 5 minutes long and the rest are 2 minutes long. What is the average (arithmetic mean) number of minutes per song?

F. 3
G. 3.3
H. 3.5
J. 4
K. 4.2

The number of songs for 3 min equals $\frac{1}{2} \times 18 = 9$ songs;
the number of songs for 5 min equals $\frac{1}{3} \times 18 = 6$ songs;
the number of songs for 2 min equals $(1 - \frac{1}{2} - \frac{1}{3}) \times 18$
$= \frac{1}{6} \times 18 = 3$ songs.
Therefore, the average number of minutes per song will be
$$\frac{(9 \times 3 \text{ min}) + (6 \times 5 \text{ min}) + (3 \times 2 \text{ min})}{18 \text{ songs}} = \frac{63 \text{ min}}{18 \text{ songs}} = \frac{7}{2} \text{ min/song.}$$

Ans. H

31. A number x can be expressed as 75 less than 125 percent of itself. What is the value of x?

A. 250 **B.** 270 **C.** 275 **D.** 300 **E.** 315

Translating mathematically, the equation becomes,
$x = (-)75 + (1.25)x$, or $75 = 0.25x$, thus, $x = 300$.

Ans. D

32. A telephone call between Irvine and Inglewood costs 50 cents for the first minute and 20 cents for each additional minute. Amanda made a 10-minute call from Irvine to Inglewood. Her brother Jerry also made a call from Irvine to Inglewood, but his call cost three times as much as Amanda's. How long was Jerry's call, in minutes? (Assume that Jerry's call was a whole number of minutes.)

F. 27 **G.** 29 **H.** 30 **J.** 32 **K.** 33

Notice that the call cost 50¢ for the first 1 minute, and 20¢ for the additional minutes. When Amanda makes 10 minutes call, the first 1 minute will cost her 50¢, whereas the remaining 9 minutes will cost 20¢. This results her cost, C = (50¢ × 1) + (20¢ × 9) = 230¢. Now, Jerry's call costs three times as much as Amanda's, or 3 × (230¢) = 690¢. Therefore, 690¢ = 50¢ × (1 minute) + 20¢ × (x minutes). Solving this eq. results in $x = 32$ minutes.
But, be careful!! You must add additional one minute for the first minute of 50¢. Therefore, $32 + 1 = 33$ minutes.

Ans. K

33. The numerals of 0, 2, 3, 5 and 7 will be used to make 4-digit numbers. If no numerals can be used more than once, how many different 4-digit numbers can be made?

A. 48
B. 56
C. 64
D. 72
E. 96

[A] [B] [C] [D]
Since it is a 4-digit numbers, [A] cannot have 0, leaving four possible numbers of 2, 3, 5 and 7. Also, [B] can be filled in with three of the rest of 2, 3, 5 and 7, and 0, leaving another four possible numbers. Remember that each number must be used only once. Therefore, [C] can be filled with the unused 3 possible numbers, leaving three possible numbers, and thus [D] with two remaining possible numbers.
Now the answer is, $4 \times 4 \times 3 \times 2 = 96$.

Ans. E

34. Steve and Tom collect stamps. If Steve had 11 more stamps, he would have 2 times as many stamps as Tom has. Steve has 135 stamps. How many stamps do the two boys have altogether?

F. 208 **G.** 235 **H.** 246 **J.** 259 **K.** 281

Let S = the number of stamps that Steve collects and T = the number of stamps that Tom collects.
Then, the equation becomes, $S + 11 = 2 \times T$.
Here, $S = 135$, and $135 + 11 = 2 \times T$. We get, $T = 73$.
Therefore, the sum of $S + T = 208$.

Ans. F

35. The 2010 minimum wage was 40 percent more than the 1990 minimum wage. If the 2010 minimum wage was \$8.75 per hour, what was the 1990 minimum wage in dollars per hour?

A. 6.25 **B.** 6.50 **C.** 6.75 **D.** 7.00 **E.** 7.25

Let x = the 1990 minimum wage, and y = the 2010 minimum wage. Then, the equation becomes, $y = (1+ 0.40)x$, since y is 40% more than x. Now, replacing y = \$8.75, we get $8.75 = (1.40)x$. Therefore, $x = \frac{8.75}{1.40} = \$6.25$.

Ans. A

36. What is the average (arithmetic mean) of all the integers from -49 to 51, inclusive?

F. -0.5
G. 0
H. 0.5
J. 1
K. 1.5

From (-)49 to 51, the sum of all integers becomes;
$(-)49 + (-)48 + (-)47 + \ldots + (-)2 + (-)1 + 0 + 1 + 2 + 3 + \ldots + 48 + 49$, and plus $50 + 51$.
Here, the sum of all (-)49 through (+)49 becomes zero, and leaving $50 + 51 = 101$ as the total sum. Also, there are 101 integers from (-)49 to 51, because it is, the last number minus the first number plus one as in $51 - (-)49 + 1 = 101$.
Therefore, the average of all integers is, $\frac{\text{sum}}{\text{no of integers}} = \frac{101}{101} = 1$.

Ans. J

37. If $x^2 - y^2 = 117$ and $x - y = 13$, what is the value of x?

A. 8 **B.** 10 **C.** 11 **D.** 13 **E.** 16

$x^2 - y^2 = (x + y)(x - y) = 117$. But $(x - y)$ is given by 13.
Replacing $x - y = 13$, we get, $(x + y)(13) = 117$.
Therefore, $x + y = 9$.
Now, solving $x + y = 9$ and $x - y = 13$, we get $x = 11$, $y = (-)2$.

Ans. C

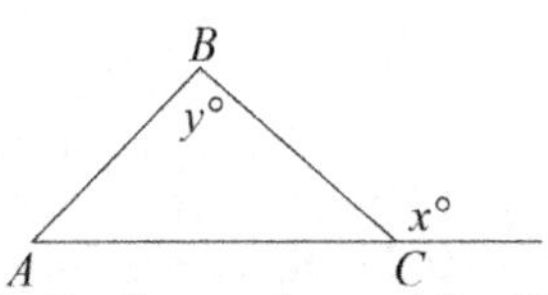

38. In the figure above, $AB = BC$ and $x = 150°$. What is the value of y?

F. 100° **G.** 110° **H.** 120° **J.** 130° **K.** 140°

Since $\overline{AB} = \overline{BC}$, we have an isosceles triangle. Therefore, the base angle $\angle A = \angle C$. But $x = 150°$, and this results $\angle C = 30°$.
Therefore, $y = 180° - \angle A - \angle C = 180° - 30° - 30° = 120°$.

Ans. H

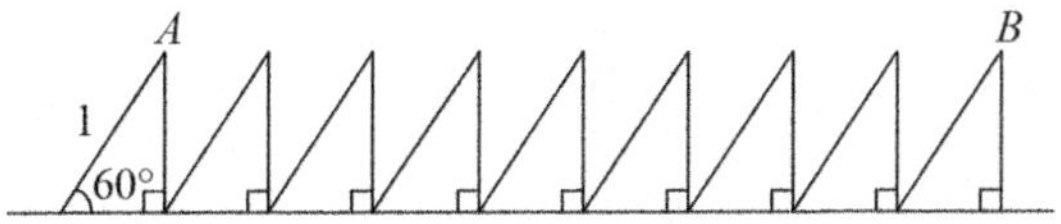

39. In the figure above, the bases of nine identical right triangles lie along a straight line. How far apart are points A and B?

A. 4
B. $4\sqrt{2}$
C. $4\sqrt{3}$
D. 5
E. It cannot be determined from the information given.

Since each triangle has 30° - 60° - 90°,
using the reference information on Special Right Triangle,
we get the corresponding side lengths in the ratio of $1:\sqrt{3}:2$.
Since the hypotenuse of the triangle is given by 1,
and the following figure, shows that the base equals $\frac{1}{2}$.

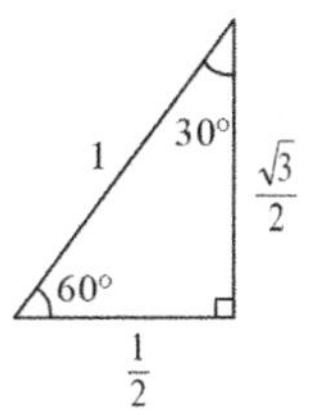

Since the length of $\overline{AB}$ equals 8 times the base length of each triangle,
we get, $\overline{AB} = 8 \times \frac{1}{2} = 4$.

Ans. A

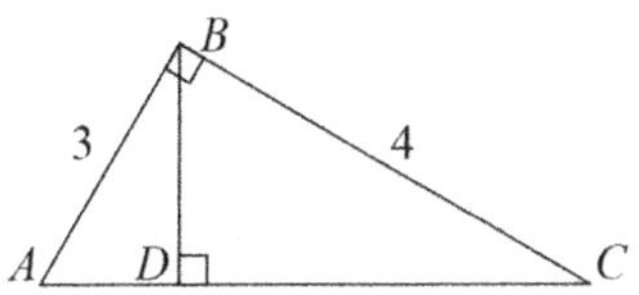

40. In the figure above, $\overline{AB} \perp \overline{BC}$, $\overline{BD} \perp \overline{AC}$, $AB = 3$ and $BC = 4$. What is the perimeter of $\triangle ABD$?

F. 7.2 **G.** 7.4 **H.** 8.6 **J.** 9.6 **K.** 12

Notice that $\triangle ABD$, $\triangle ACB$ and $\triangle CBD$ are all similar to each other, because of angle-angle-angle theorem. Also, $\triangle ACB$ is a special right triangle with its side length, 3-4-5, or $AC = 5$.
Using this fact, $\frac{\overline{AC}}{\overline{CB}} = \frac{\overline{AB}}{\overline{BD}}$, or $\frac{5}{4} = \frac{3}{\overline{BD}}$.
Therefore, we get $\overline{BD} = \frac{12}{5} = 2.4$.

Ans. F

41. At his normal swimming speed, Eugene swims a distance of d meters in 4 minutes. If he swims only at $\frac{2}{3}$ of his normal swimming speed, how many minutes will it take him to swim a distance of $\frac{d}{2}$ meters?

A. $\frac{4}{3}$ **B.** 2 **C.** $\frac{8}{3}$ **D.** 3 **E.** $\frac{10}{3}$

Referring to the formula, Speed $= \frac{\text{Distance}}{\text{Time}}$,
we get Eugene's speed, $S = \frac{d}{4}$. But if he swims at $\frac{2}{3}$ of this speed,
his new speed will be $\frac{2}{3} \times S = \frac{2}{3} \times \frac{d}{4} = \frac{d}{6}$.
Now with this new speed, we want to get time for a distance of $\frac{d}{2}$ meters. Using the fact, Time $= \frac{\text{Distance}}{\text{Speed}}$, we get $T = \frac{\frac{d}{2}}{\frac{d}{6}} = 3$.

Ans. D

42. Eight points in a plane are arranged so that no 3 of them are collinear. What is the total number of straight lines connecting two points?

F. 24 **G.** 28 **H.** 12 **J.** 16 **K.** 18

Here, we may use the combination formula, ${}_nC_r$, where n is the total number of points, and r is the number points that we choose.
In our problem, $n = 8$ total points, whereas $r = 2$ points that we choose to draw a line. Therefore, ${}_8C_2 = \frac{8!}{6!2!} = \frac{8 \times 7}{2} = 28$.

Ans. G

$a < b < c$
$d < e < f$
$e < a$
$f < c$

43. If each letter in the inequalities above represents a positive integer, then, of the following fractions, which is the smallest?

A. $\frac{a}{e}$ **B.** $\frac{d}{8}$ **C.** $\frac{d}{b}$ **D.** $\frac{c}{f}$ **E.** $\frac{d}{c}$

First of all,
from $a < b < c$ and $e < a$, we get $e < a < b < c$ – ①
Also from $d < e < f$ and $f < c$, we get $d < e < f < c$ – ②
Considering these two inequalities,
$e < a < b < c$ – ①
$d < e < f < c$ – ②
We have $d < e$, thus d is the smallest, while c is the greatest.
Therefore, the smallest fraction will be $\frac{\text{the smallest}}{\text{the greatest}} = \frac{d}{c}$.

Ans. E

44. If a nonzero number n is divided by its reciprocal, what is the result?

F. 1 **G.** n **H.** $2n$ **J.** n^2 **K.** $2n^2$

Translating mathematically, we get $n \div \frac{1}{n} = n \times \frac{n}{1} = n^2$.

Ans. J

45. If 20 percent of 60 percent of a positive number is equal to w percent of the same number, what is the value of w?

A. 1,200 **B.** 120 **C.** 12 **D.** 1.2 **E.** 0.12

Mathematical translation gives, $(0.20) \times (0.60) \times n = \frac{w}{100} \times n$.
By solving the eq., we get $w = 100 \times (0.20) \times (0.60) = 12$.
Ans. C

46. During a sale, a customer can buy one shirt for x dollars. The second shirt and the rest of the shirts that the customer buys is discounted to $\frac{x}{2}$ dollars. For example, the cost of 5 shirts is $3x$ dollars. Which of the following represents the customer's cost, in dollars, for $2n$ shirts bought during this sale?

F. $\left(\frac{2n - 3}{2}\right)x$ **G.** $\left(\frac{2n - 1}{2}\right)x$ **H.** $(2n)x$

J. $(2n - 1)x$ **K.** $\left(\frac{2n + 1}{2}\right)x$

For $2n$ shirts, we pay the first shirt at x dollars, the second shirt at $\frac{x}{2}$ dollars, the third shirt at $\frac{x}{2}$ dollars… the $(2n)$th shirt at $\frac{x}{2}$ dollars. Out of total $2n$ shirts, we have already paid x dollar for the first one shirt, leaving the rest of $(2n - 1)$ shirts at $\frac{x}{2}$ dollars.
Therefore, the total cost C is,

$$C = (1)(x) + (2n - 1)\frac{x}{2} = \left(\frac{2n + 1}{2}\right)x.$$

Ans. K

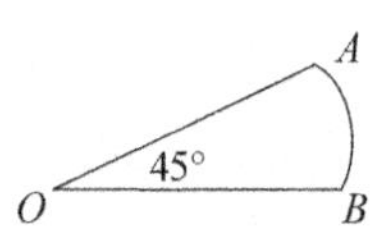

47. In the figure above, $\overset{\frown}{AB}$ is the arc of a circle with center O. If the length of arc $\overset{\frown}{AB}$ is 3π, what is the area of region OAB?

A. 45π
B. 36π
C. 27π
D. 18π
E. 9π

Remember the formula for the arc length, $\overset{\frown}{AB} = r \times \theta$, where θ is an angle in radian mode, we get $\overset{\frown}{AB} = r \times \theta = r \times \frac{\pi}{4} = 3\pi$ -- ①
Here, r is the radius $\overline{OA}$, and $\theta = 45° = \frac{\pi}{4}$ in radian mode.

(Note: $45° = 45° \times \frac{\pi}{180°} = \frac{\pi}{4}$ rad.)
Now, from eq. ①, we get $r = 12$. Also, notice that the area of OAB is, area $= \frac{1}{2} r^2 \times \theta$, where θ is in radian mode.

Therefore, area $OAB = \frac{1}{2} \times (12)^2 \times \frac{\pi}{4} = 18\pi$.

Ans. D

(Please note that how the arc length $\widehat{AB}$ and the area formula have been converted from Degree mode to Radian mode:

$\widehat{AB} = 2\pi r \times \frac{\theta}{360^\circ}$, but $360^\circ = 2\pi$ rad.

Thus $\widehat{AB} = 2\pi r \times \frac{\theta}{2\pi} = r \times \theta$. For the area OAB,

say A, $A = \pi r^2 \times \frac{\theta}{360^\circ} = \pi r^2 \times \frac{\theta}{2\pi} = \frac{1}{2} r^2 \times \theta$)

48. A water tank measuring 3 yards by 4 yards by 2 yards is filled to capacity by a pipe in $\frac{1}{3}$ hour. How many hours will it take to fill a water tank measuring 6 yards by 8 yards by 3 yards to capacity, if it is filled by the same pipe with water flowing at the same rate?

F. 1 **G.** 2 **H.** 4 **J.** 8 **K.** 13.5

The speed of the pipe is

$S = \frac{\text{Volume}}{\text{Time}}$ or $S = \frac{(3 \times 4 \times 2)\text{ yd}^3}{\frac{1}{3}\text{ hr.}} = \frac{72\text{ yd}^3}{\text{hr.}}$.

Now, with this speed, we want to get time, Time $= \frac{\text{Volume}}{\text{Speed}}$.

Our new volume, $V = 6 \times 8 \times 3$, and speed $S = 72$.

Therefore, $T = \frac{(6 \times 8 \times 3)}{72} = 2$ hrs.

Ans. G

49. Five distinct lines in a plane are such that no two lines are parallel and no three lines intersect at the same point. What is the total number of points of intersection of these five lines?

A. Four
B. Five
C. Six
D. Eight
E. Ten

Refer to the following figure that is drawn according to the instruction given in this problem. Here, we get a total of 10 intersection points. Or, alternatively, we may use combination formula ${}_5C_2 = \frac{5!}{3!2!} = 10$, because we choose 2 lines out of 5, to get an intersection point.

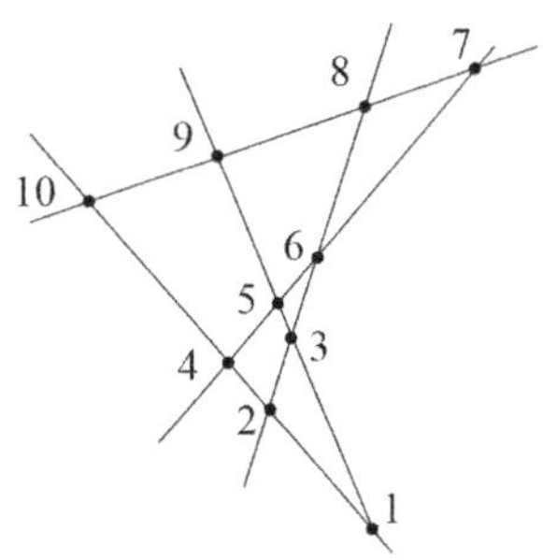

Ans. E

50. Jim is working t hours each day and spends $\frac{t}{5}$ hours each day for a marketing project. In terms of t, how many days will it take Jim to complete the project that require a total of 20 hours?

F. $\frac{t}{100}$ **G.** $\frac{t}{20}$ **H.** $\frac{20}{t}$ **J.** $\frac{100}{t}$ **K.** $20t$

He spends $\frac{t}{5}$ hrs. to work on his project per day.

To get the total number of days to finish the project will be,

$\frac{\text{(total hours)}}{\text{(number of hrs. per day)}}$, or $\frac{\text{(20 hrs.)}}{(\frac{t}{5}\text{ per day})} = \frac{100}{t}$ days.

Ans. J

51. If $t > 0$, which of the following is equivalent to 25 percent of t?

A. $t - 0.75$ **B.** $t - 0.25$ **C.** $\frac{t}{25}$ **D.** $\frac{t}{75}$ **E.** $\frac{t}{4}$

$(0.25) \times t = \frac{1}{4} \times t = \frac{t}{4}$

Ans. E

52. A positive integer is called a palindrome if it reads the same forward as it does backward. For example, 959 and 8228 are palindromes, whereas 1332 is not. Neither the first nor the last digit of palindrome can be 0. How many palindromes are there between 100 and 1,500?

F. 94 **G.** 95 **H.** 104 **J.** 105 **K.** 1,400

A palindrome is a positive integer with a symmetric figure. Therefore, between 100 and 1,500, we have
101, 111, 121, ... , 191
202, 212, 222, ... , 292
...
909, 919, 929, ... , 999, and up to here,
we have total of $9 \times (10) = 90$ integer.
To continue, we have 5 more integers as in 1001, 1111, 1221, 1331, and 1441. Therefore, we get total of 95 integers.

Ans. G

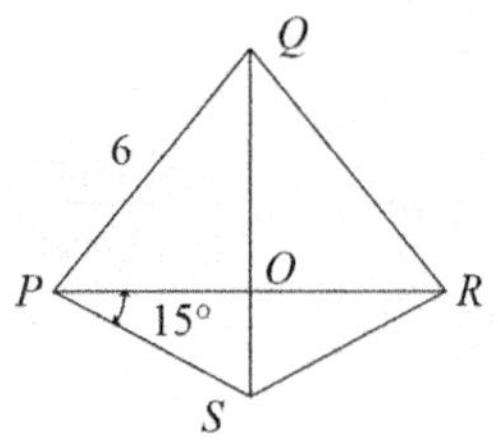

53. In the quadrilateral above, if $PQ = SQ = RQ$, $PS = SR$ and $PQ = 6$, then $PR =$

A. $2\sqrt{3}$ **B.** $3\sqrt{2}$ **C.** 4 **D.** 5 **E.** 6

First, consider $\triangle PSO$.
Since the quadrilateral *PSRQ* is a kite with each two sides have the same lengths, we know that its diagonals are perpendicular to each other.
Here, $\angle POS = 90°$, which leaves $\angle PSQ = 75°$.
Also, $\triangle PQS$ is an isosceles triangle with the same base angles, we get $\angle SPQ = 75°$ also.
This leads us to get $\angle QPO = 75° - 15° = 60°$.
Now, consider $\triangle PQO$ is a special right triangle with 30°-60°-90° with its side length ratios of $1:\sqrt{3}:2$.
Since we have the hypotenuse $\overline{PQ} = 6$, we get $\overline{PO} = 3$.
Therefore, $\overline{PR} = 2 \times \overline{PO} = 2 \times 3 = 6$.

Ans. E

54. Stack of 20 boxes of equal height is 4 feet high. What is the height, in inches, of each box?

F. 4.8 **G.** 2.4 **H.** 3.6 **J.** 1.2 **K.** 2

Since each box has the same height,
and the total height of the stack is 4 ft., or $4 \times (12 \text{ in.}) = 48$ in.,
we get the height of each box, $\frac{(48 \text{ in.})}{(20 \text{ boxes})} = (2.4 \text{ inches})$.

Ans. G

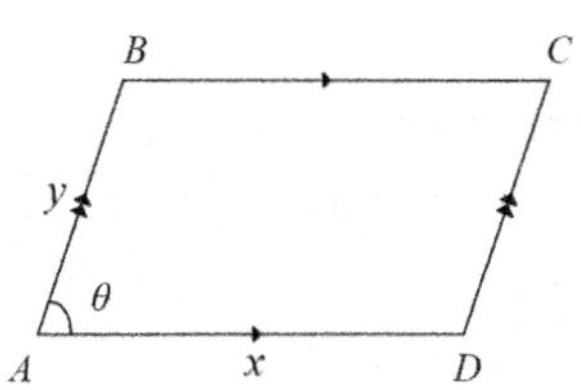

55. What is the area of the parallelogram *ABCD* in above figure, in terms of x, y, and θ?

A. $\frac{1}{2}xy$ **B.** xy **C.** $xy \cos\theta$ **D.** $xy \sin\theta$ **E.** $xy \tan\theta$

The area is, Base × Height $= x \times (y \sin\theta) = xy \sin\theta$

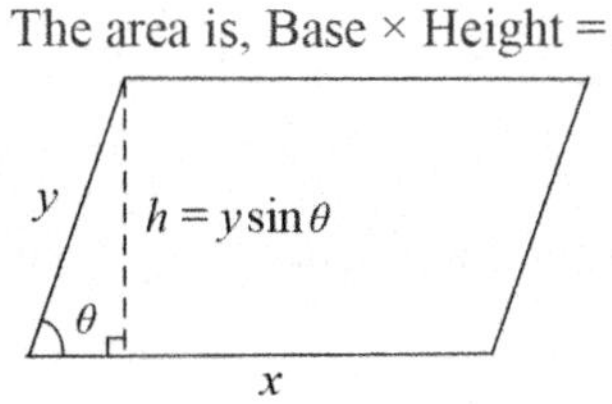

Ans. D

56. If the graph of $f(x) = x^2 + 1$ is translated 3 units to the right and then 4 units down, what would be the value of the transformed graph at the point $x = 4$?

F. 1.5 **G.** 5.1 **H.** -7.2 **J.** -2.0 **K.** -2.5

$f(x) = x^2 + 1$
Since 3 units to the right, we get $f(x - 3) = (x - 3)^2 + 1$.
And then, 4 units down, we get $f(x - 3) - 4 = (x - 3)^2 + 1 - 4$.
Therefore, new function $g(x) = (x - 3)^2 - 3$,
and $g(4) = (4 - 3)^2 - 3 = -2$

Ans. J

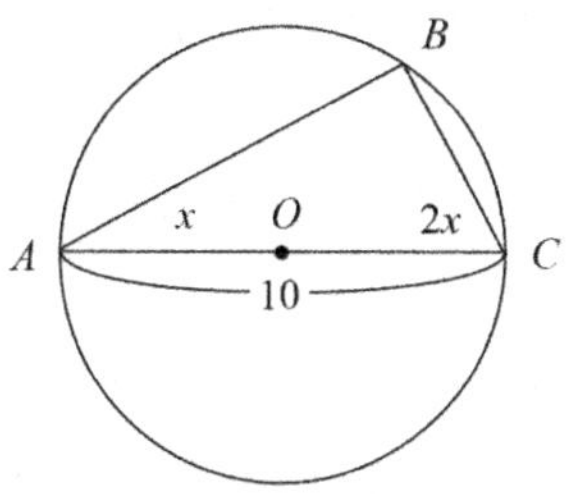

57. In above figure, what is the area of ΔABC?

A. 21.65 **B.** 20.21 **C.** 20.19
D. 20.17 **E.** 18.01

First of all, the sum of all interior angles is, $x + 2x + 90° = 180°$
$\therefore x = 30°$
Therefore, it is 30°-60°-90° right triangle.
Since the hypotenuse $\overline{AC} = 10$, $\overline{BC}$ becomes 5 and $\overline{AB} = 5\sqrt{3}$.
So, the area is $\frac{1}{2}(5\sqrt{3})(5) = 21.65$

Ans. A

58. If $x_0 = 1$ and $x_{n+1} = (x_n - 1)^2$, what is the value of x_3?

F. -3 **G.** -2 **H.** -1 **J.** 0 **K.** 1

$x_1 = (x_0 - 1)^2 = (1 - 1)^2 = 0$
$x_2 = (x_1 - 1)^2 = (0 - 1)^2 = 1$
$x_3 = (x_2 - 1)^2 = (1 - 1)^2 = 0$
$\therefore x_3 = 0$

Ans. J

59. If $p(x) = ax^3 + bx^2 + cx + d$ such that $p(0) = 1$, $p(1) = 3$, and $p(-1) = 5$, then the value of b is

A. 3 **B.** 2 **C.** 1 **D.** 0
E. Not enough information to tell

$p(0) = 0 + 0 + 0 + d = 1 \quad \therefore d = 1$
$p(1) = a + b + c + d = 3$
$p(-1) = -a + b - c + d = 5$
Therefore, $p(1) + p(-1) = 2b + 2d = 8$
$\therefore b + d = 4$, but $d = 1$
$\therefore b = 3$

Ans. A

60. If $f(x) = \frac{1}{1 + x^2}$ and $g(x) = \sqrt{x - 4}$, then what is the value of $f(g(13))$?

F. $\frac{1}{4}$ **G.** $\frac{1}{8}$ **H.** $\frac{2}{3}$ **J.** $\frac{1}{5}$ **K.** $\frac{1}{10}$

First of all, $g(13) = \sqrt{13 - 4} = \sqrt{9} = 3$
Therefore, $f(g(13)) = f(3) = \frac{1}{1 + 3^2} = \frac{1}{10}$

Ans. K

Model Test No. 02

60 Questions / 60 Minutes

Directions: Solve each problem, choose the correct answer, and then fill in the corresponding oval on your answer document.

Do not linger over problems that take too much time. Solve as many as you can; then return to the others in the time you have left for this test.

You are permitted to use a calculator on this test. You many use your calculator for any problems you choose, but some of the problems may best be done without using a calculator.

Note: Unless otherwise stated, all of the following should be assumed.

1. Illustrative figures are NOT necessarily drawn to scale.
2. Geometric figures lie in a plane.
3. The word line indicates a straight line.
4. The word average indicates arithmetic mean.

Answer Sheet

Model Test No. 02

1	Ⓐ Ⓑ Ⓒ Ⓓ Ⓔ	**16**	Ⓕ Ⓖ Ⓗ Ⓙ Ⓚ	**31**	Ⓐ Ⓑ Ⓒ Ⓓ Ⓔ	**46**	Ⓕ Ⓖ Ⓗ Ⓙ Ⓚ
2	Ⓕ Ⓖ Ⓗ Ⓙ Ⓚ	**17**	Ⓐ Ⓑ Ⓒ Ⓓ Ⓔ	**32**	Ⓕ Ⓖ Ⓗ Ⓙ Ⓚ	**47**	Ⓐ Ⓑ Ⓒ Ⓓ Ⓔ
3	Ⓐ Ⓑ Ⓒ Ⓓ Ⓔ	**18**	Ⓕ Ⓖ Ⓗ Ⓙ Ⓚ	**33**	Ⓐ Ⓑ Ⓒ Ⓓ Ⓔ	**48**	Ⓕ Ⓖ Ⓗ Ⓙ Ⓚ
4	Ⓕ Ⓖ Ⓗ Ⓙ Ⓚ	**19**	Ⓐ Ⓑ Ⓒ Ⓓ Ⓔ	**34**	Ⓕ Ⓖ Ⓗ Ⓙ Ⓚ	**49**	Ⓐ Ⓑ Ⓒ Ⓓ Ⓔ
5	Ⓐ Ⓑ Ⓒ Ⓓ Ⓔ	**20**	Ⓕ Ⓖ Ⓗ Ⓙ Ⓚ	**35**	Ⓐ Ⓑ Ⓒ Ⓓ Ⓔ	**50**	Ⓕ Ⓖ Ⓗ Ⓙ Ⓚ
6	Ⓕ Ⓖ Ⓗ Ⓙ Ⓚ	**21**	Ⓐ Ⓑ Ⓒ Ⓓ Ⓔ	**36**	Ⓕ Ⓖ Ⓗ Ⓙ Ⓚ	**51**	Ⓐ Ⓑ Ⓒ Ⓓ Ⓔ
7	Ⓐ Ⓑ Ⓒ Ⓓ Ⓔ	**22**	Ⓕ Ⓖ Ⓗ Ⓙ Ⓚ	**37**	Ⓐ Ⓑ Ⓒ Ⓓ Ⓔ	**52**	Ⓕ Ⓖ Ⓗ Ⓙ Ⓚ
8	Ⓕ Ⓖ Ⓗ Ⓙ Ⓚ	**23**	Ⓐ Ⓑ Ⓒ Ⓓ Ⓔ	**38**	Ⓕ Ⓖ Ⓗ Ⓙ Ⓚ	**53**	Ⓐ Ⓑ Ⓒ Ⓓ Ⓔ
9	Ⓐ Ⓑ Ⓒ Ⓓ Ⓔ	**24**	Ⓕ Ⓖ Ⓗ Ⓙ Ⓚ	**39**	Ⓐ Ⓑ Ⓒ Ⓓ Ⓔ	**54**	Ⓕ Ⓖ Ⓗ Ⓙ Ⓚ
10	Ⓕ Ⓖ Ⓗ Ⓙ Ⓚ	**25**	Ⓐ Ⓑ Ⓒ Ⓓ Ⓔ	**40**	Ⓕ Ⓖ Ⓗ Ⓙ Ⓚ	**55**	Ⓐ Ⓑ Ⓒ Ⓓ Ⓔ
11	Ⓐ Ⓑ Ⓒ Ⓓ Ⓔ	**26**	Ⓕ Ⓖ Ⓗ Ⓙ Ⓚ	**41**	Ⓐ Ⓑ Ⓒ Ⓓ Ⓔ	**56**	Ⓕ Ⓖ Ⓗ Ⓙ Ⓚ
12	Ⓕ Ⓖ Ⓗ Ⓙ Ⓚ	**27**	Ⓐ Ⓑ Ⓒ Ⓓ Ⓔ	**42**	Ⓕ Ⓖ Ⓗ Ⓙ Ⓚ	**57**	Ⓐ Ⓑ Ⓒ Ⓓ Ⓔ
13	Ⓐ Ⓑ Ⓒ Ⓓ Ⓔ	**28**	Ⓕ Ⓖ Ⓗ Ⓙ Ⓚ	**43**	Ⓐ Ⓑ Ⓒ Ⓓ Ⓔ	**58**	Ⓕ Ⓖ Ⓗ Ⓙ Ⓚ
14	Ⓕ Ⓖ Ⓗ Ⓙ Ⓚ	**29**	Ⓐ Ⓑ Ⓒ Ⓓ Ⓔ	**44**	Ⓕ Ⓖ Ⓗ Ⓙ Ⓚ	**59**	Ⓐ Ⓑ Ⓒ Ⓓ Ⓔ
15	Ⓐ Ⓑ Ⓒ Ⓓ Ⓔ	**30**	Ⓕ Ⓖ Ⓗ Ⓙ Ⓚ	**45**	Ⓐ Ⓑ Ⓒ Ⓓ Ⓔ	**60**	Ⓕ Ⓖ Ⓗ Ⓙ Ⓚ

MATHEMATICS TEST

60 Minutes – 60 Questions

DIRECTIONS: Solve each problem, choose the correct answer, and then fill in the corresponding oval on your answer document.

Do not linger over problems that take too much time. Solve as many as you can; then return to the others in the time you have left for this test.

You are permitted to use a calculator on this test. You many use your calculator for any problems you choose, but some of the problems may best be done without using a calculator.

Note: Unless otherwise stated, all of the following should be assumed.

1. Illustrative figures are NOT necessarily drawn to scale.
2. Geometric figures lie in a plane.
3. The word line indicates a straight line.
4. The word average indicates arithmetic mean.

---

1. Auto Sales leases a new automobile for a down payment of \$3,200 plus \$480 per month for 36 months. Which of the following functions models the total amount, in dollars, at the end of m months of the lease?

A. $f(m) = 3{,}200 + m$
B. $f(m) = 3{,}200 + 36m$
C. $f(m) = 3{,}200 + 480m$
D. $f(m) = 3{,}200 + 480(36 - m)$
E. $f(m) = 3{,}680 + 480(m - 1)$

2. If $y = x^2 - 2x - 1$ and x is a real number, which of the following CANNOT be the value of y?

F. -2
G. -1
H. 0
J. 1
K. -5

x	$f(x)$
-1	1
-2	-2
-3	4
-5	6
-6	5

3. The function f is defined by the table above. For which value of x does $f(x) = 5$?

A. -1
B. -2
C. -3
D. -5
E. -6

4. If $x^2 + 4x \leq -3$, which of the following is a possible value of x?

F. -6
G. -4
H. -2
J. 0
K. 2

5. If $2n + p = 5$ and $2m - 4n = 2$, what is the value of $m + p$?

A. 6
B. 4
C. 5
D. 7
E. 8

6. If $3^y = z$, which of the following must be equal to 3^{y-1}?

F. $\frac{z}{3}$
G. $z + 3$
H. $3z$
J. $\frac{z}{9}$
K. $\frac{z^2}{3}$

7. If n is a two-digit number and the sum of its units digit and tens digit is 9, which of the following statements must be true?

A. n is a multiple of 4.
B. n is a multiple of 5.
C. n is a multiple of 6.
D. n is a multiple of 9.
E. n is a multiple of 13.

GO ON TO THE NEXT PAGE.

8. 10 percent of 50 is 50 percent of what number?

F. 40
G. 20
H. 10
J. 4
K. 2

$k^2 \geq k^3$ for all real numbers k

9. Which of the following values for k shows that the statement above is false?

A. -1
B. 0
C. $\frac{3}{2}$
D. 1
E. $\frac{1}{2}$

10. If k is 10 percent of p and p is 20 percent of s, what percent of s is k?

F. 2%
G. 4%
H. 10%
J. 40%
K. 100%

11. A certain restaurant parlor offers 8 different toppings for pizzas. How many combinations of 3 different toppings from those offered are possible?

A. 6
B. 12
C. 15
D. 36
E. 56

$$\frac{4}{x-2} > 1$$

12. In which of the following intervals does every value of x satisfy the inequality above?

F. $-6 \leq x \leq -5$
G. $-5 \leq x \leq -3$
H. $3 \leq x \leq 5$
J. $2 \leq x \leq 6$
K. $6 \leq x \leq 8$

x	y
$\frac{1}{6}$	3
$\frac{1}{3}$	6
$\frac{5}{9}$	?
2	36
3	54

13. In the table above, x and y are directly proportional. What is the value of y when x is $\frac{5}{9}$?

A. 8
B. 9
C. 10
D. 12
E. 18

1, 2, 4, 8 …

14. The first term of the sequence above is 1, and each term after the first is twice the preceding term. Which of the following expressions represents the nth term of the sequence?

F. 2^n
G. $2^n - 1$
H. 2^{n-1}
J. $(n-1)^2$
K. $2(n-1)$

Alice's Monthly Budget

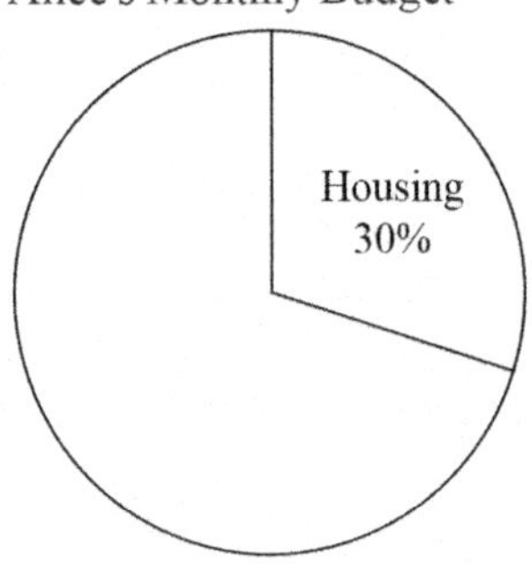

15. On the basis of the information in the graph above, if Alice's monthly housing budget is $900, what is the dollar amount of her total monthly budget?

A. $3,000
B. $2,700
C. $3,600
D. $1,800
E. $900

GO ON TO THE NEXT PAGE.

16. When 10^3 is added to 34,567, the value of which digit in the number 34,567 will be increased?

F. 3
G. 4
H. 5
J. 6
K. 7

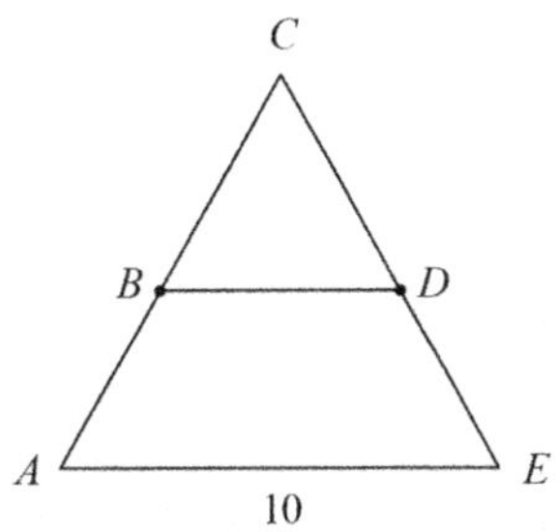

Note: Figure not drawn to scale.

17. In $\triangle ACE$ above, $AC = 2BC$ and $CE = 2CD$. If $AC = 9$ and $CE = 15$, what is the perimeter of $\triangle BCD$?

A. 13
B. 14
C. 15
D. 16
E. 17

$N(p) = -1.5p + 45$

18. For a certain product, the number N of thousands of units sold each month is given by the function above, where p is the price per unit, in dollars, and $0 < p \leq 10$. The price per unit was \$10 in July and \$8 in August. How does the number of units sold in July compare to the number of units sold in August?

F. 3,000 fewer units were sold in July.
G. 8,000 fewer units were sold in July.
H. 20,000 fewer units were sold in July.
J. 12,000 more units were sold in July.
K. 16,000 more units were sold in July.

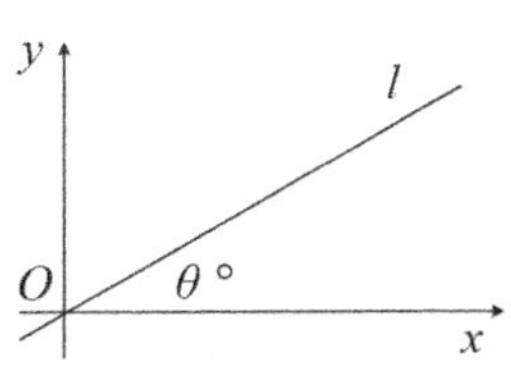

19. In the figure above, what is the equation of line l?

A. $y = x \sin\theta$
B. $y = x \cos\theta$
C. $y = x \tan\theta$
D. $y = x \csc\theta$
E. $y = x \sec\theta$

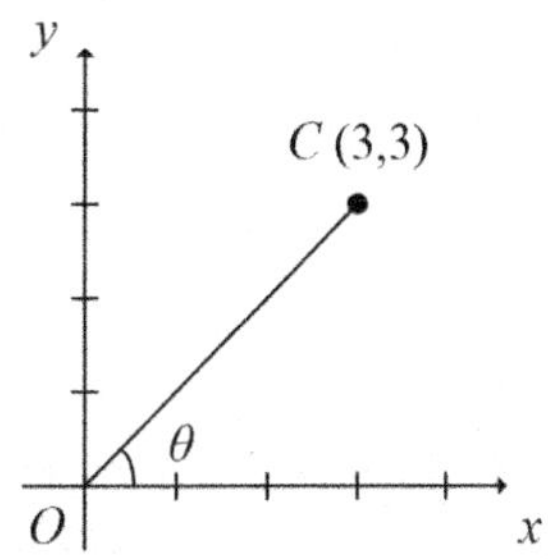

20. In the xy-coordinate system above, what is the angle θ formed by $\overline{OC}$ and x-axis?

F. 36°
G. 18°
H. 30°
J. 60°
K. 45°

$|x - 5| < 3$
$|y + 2| < 7$

21. If x and y satisfy the inequalities above, then $|xy|$ could be any of the following EXCEPT

A. 18
B. 56
C. 64
D. 62
E. 80

22. If the diagonals of quadrilateral P are perpendicular to each other, which of the following statements must be true?

I. All sides of P have equal length.
II. Each side of P is parallel to the opposite side.
III. The measure of each interior angle of P is 90°

F. None
G. I only
H. II only
J. III only
K. I, II, and III

23. The first term of a sequence is 3 and the second term is 1. From the third term on, each term is the sum of all preceding terms. What is the 10th term of the sequence?

A. 512
B. 768
C. 1024
D. 1280
E. 2048

GO ON TO THE NEXT PAGE.

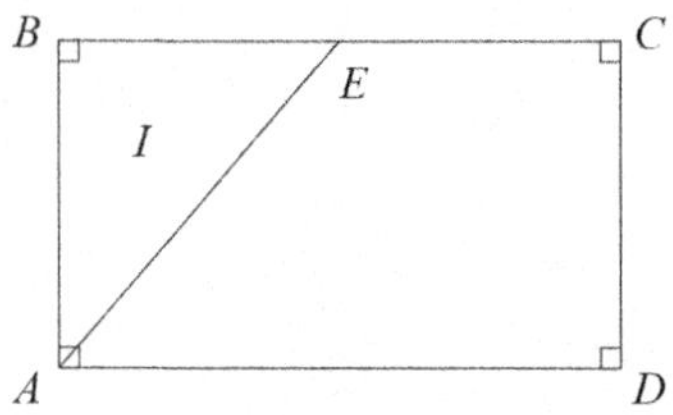

24. In the figure above, the area of triangle I is $\frac{1}{4}$ the area of rectangle $ABCD$, and the area of the triangle is 8. If the length of AB is equal to the length of BE, then what is the length of AD?

F. 8
G. 16
H. 12
J. 10
K. 18

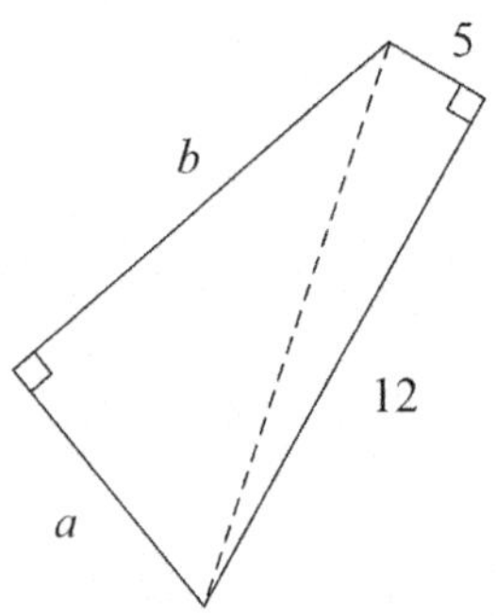

25. In the figure above, what is the value of $a^2 + b^2$?

A. 169
B. 144
C. 124
D. 256
E. 162

26. In a race, Max and Tom started at the same time and ran with constant speeds of 10 kilometers and 16 kilometers per hour, respectively. If Max crossed the finish line 22.5 minute after Tom, how many kilometers long was the race?

F. 0.25
G. 0.5
H. 2.0
J. 10.0
K. 30.0

27. If $x = t + 3$ and $y = 2t - 5$, which of the following is an expression for y, in terms of x?

A. $2x$
B. $2x + 11$
C. $2x - 11$
D. $2x - 1$
E. $2x + 1$

$$\begin{array}{r} \text{AD} \\ +\,\text{AV} \\ \hline \text{DVD} \end{array}$$

28. In the correctly worked addition problem above, A, D and V represent different digits. What is the value of A?

F. 0
G. 1
H. 5
J. 8
K. 9

29. If $(7 \times 10^5) + (4 \times 10^n) = (7.4 \times 10^5)$, what is the value of n?

A. 3
B. 4
C. 6
D. 7
E. 9

30. A large solid cube was assembled by gluing together small cubic blocks. All six faces of the large cube were then painted black. When the large cube was disassembled, 8 of the small cubic blocks had no black paint on their faces. How many small cubic blocks had black paint on any of their faces?

F. 36
G. 42
H. 48
J. 56
K. 64

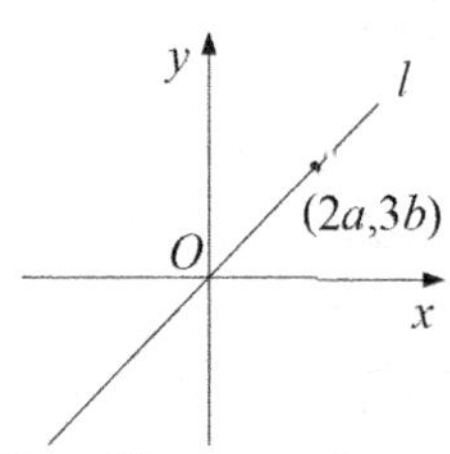

Note: Figure not drawn to scale.

31. In the figure above, if $a = 4$ and $b = 2$, what is the slope of line l?

A. $\frac{3}{8}$
B. $\frac{4}{9}$
C. $\frac{1}{2}$
D. $\frac{5}{8}$
E. $\frac{3}{4}$

GO ON TO THE NEXT PAGE.

32. Points A and B are on a circle with center O. If the measure of angle AOB is 2 radian and $AO = 5$, what is the length of minor arc AB?

F. 9.78
G. 10
H. 10.34
J. 11
K. 11.16

33. A circle with radius r has three non-overlapping arcs with lengths 3π, 5π, and 9π. If the radius is given by $r = 10$, then what is the remaining arc length that makes up the entire circle?

A. 2π
B. 4π
C. 3π
D. π
E. 5π

34. If 8 numbers are paired so that the four pairs of numbers have averages (arithmetic means) of 9, 13, 17 and 21, respectively, what is the sum of the 8 numbers?

F. 114
G. 115
H. 118
J. 120
K. 121

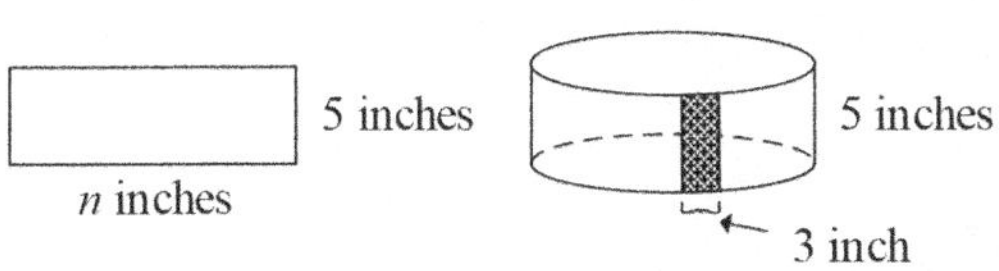

Figure I Figure II

Note: Figures not drawn to scale.

35. The rectangular sheet of paper shown in Figure I above is rolled into the shape of a right circular cylinder, as shown in Figure II, The shaded region of width 3 inch shows where the two ends of the paper overlap. If the volume of the cylinder is $\frac{125}{\pi}$ cubic inches, what is the value of n?

A. 12
B. 13
C. 15
D. 19
E. 22

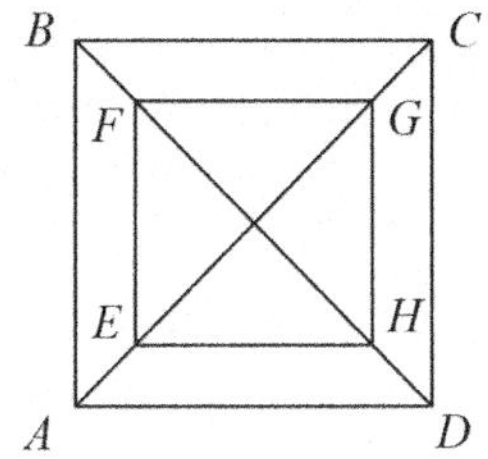

Note: Figure not drawn to scale.

36. If the area of square $ABCD$ is 144, and $AE = BF = CG = DH = 3\sqrt{2}$, what is the perimeter of the square $EFGH$?

F. 24
G. 25
H. 28
J. 30
K. 31

37. Jim cut a circular pie into wedge-shaped pieces. The tip of each piece at the center of the pie and the angle at the tip is always greater than 25°, but less than 30°. What is the least value for the number of pieces into which the pie is cut?

A. 10
B. 12
C. 13
D. 15
E. 16

38. If $(x + 5)(x - 5) = 75$, what is the value of x^2?

F. 88
G. 94
H. 98
J. 100
K. 104

39. On a bus trip to Seattle, all of the seats were occupied and 3 people were standing. On the return trip, 11 of the seats were empty and no one was standing. If the total bus fares were \$255 and \$185, respectively, for the two trips and all passengers paid the same fare, what was the passenger seating capacity of the bus?
(Assume one passenger per seat.)

A. 45
B. 48
C. 53
D. 55
E. 60

GO ON TO THE NEXT PAGE.

40. James has to sell a box of candy bars for a fund-raiser. Yesterday he sold $\frac{2}{5}$ of the candy bars in the box, and today he sold 16 candy bars. If his sales for these two days was $\frac{2}{3}$ of the total number of candy bars that came in the box, how many candy bars are left in the box?

F. 20
G. 30
H. 40
J. 60
K. 90

41. If k is a positive integer, then $(-1)^k(-1)^{k+1} =$

A. -2
B. -1
C. 1
D. $-2k - 1$
E. $\frac{1}{2k+1}$

42. The balls in a certain basket are either red or blue. If the ratio of red balls to blue balls is $\frac{1}{5}$, each of the following could be the number of balls in the basket EXCEPT

F. 15
G. 30
H. 42
J. 60
K. 90

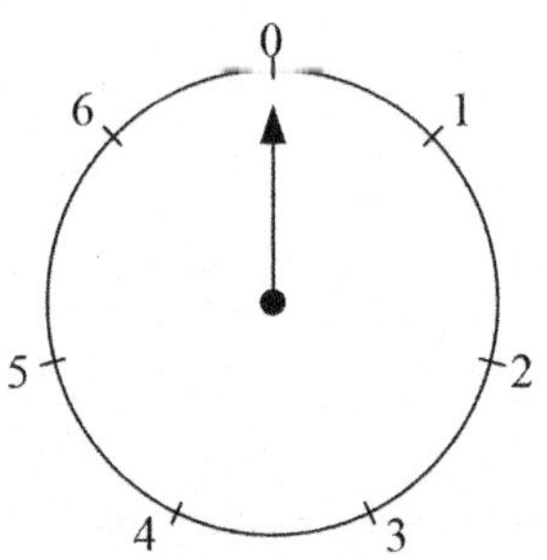

43. On the dial shown above, the numbers are equally spaced. The arrow rotates clockwise at a constant rate through an angle of 100° every 10 seconds. If the arrow starts at 0, how many times will the arrow pass the number 2 on the dial in 2 minutes?

A. Two
B. Three
C. Four
D. Five
E. Six

44. If x is a positive integer and if $y = x + \frac{1}{x}$, which of the following must be true?

I. $y \geq 2$
II. $xy \geq 1$
III. $xy > x^2$

F. I only
G. III only
H. I and II only
J. I and III only
K. I, II, and III

45. Points A, B, and C are on line l in that order. P is a point not on l. The measure of $\angle ABP$ is $(6x - 15)°$ and the measure of $\angle PBC$ is $(2x + 16)°$. What is the value of x?

A. 11.75
B. 21.5
C. 22.4
D. 23
E. 23.5

46. Which of the following is one possible value of k for which $k < 3 < \frac{1}{k}$?

F. $\frac{1}{4}$
G. $\frac{1}{2}$
H. $\frac{1}{3}$
J. 2
K. 4

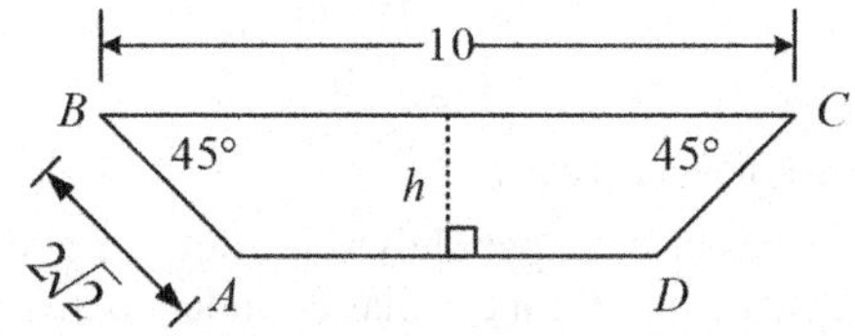

Note: Figure not drawn to scale.

47. In the figure above, if $\overline{BC} \parallel \overline{AD}$ and $\angle ABC = \angle DCB$, what is the area of $ABCD$?

A. 14
B. 15
C. 16
D. 17
E. 18

GO ON TO THE NEXT PAGE.

48. Brad and Patrick can each mow the lawn in an hour, and their father can mow the lawn in 30 minutes. If all three work together at their respective rates, how many minutes will it take to mow the lawn?

F. 5
G. 10
H. 15
J. 25
K. 30

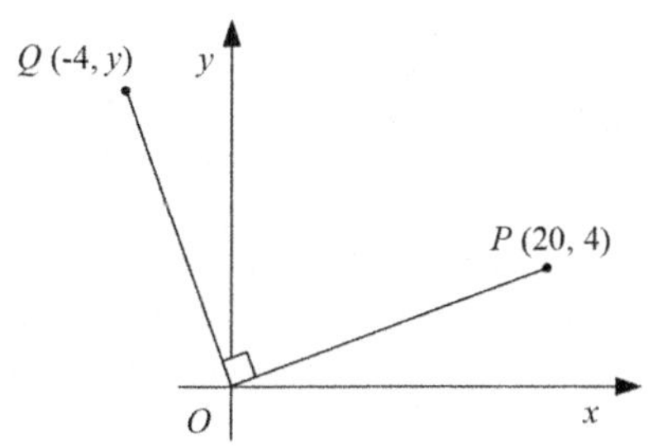

49. In the xy-plane above, $OQ = OP$. What is the value of y?

A. 20
B. 10
C. 14
D. 12
E. 8

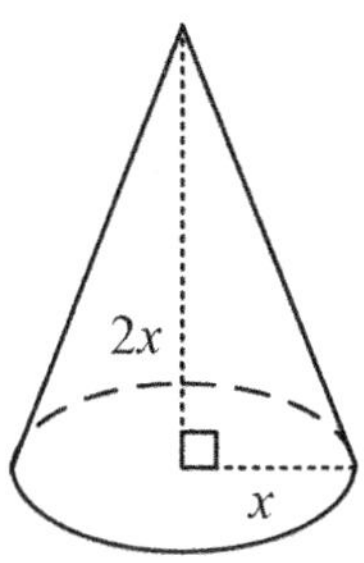

50. Which of the following has the same volume as the cone shown above with radius x and height $2x$?

F. A circular cylinder with radius $\frac{x}{3}$ and height $6x$

G. A circular cylinder with radius $\frac{2\pi x}{3}$ and height $3x$

H. A cube with edge $2x$

J. A cube with edge $2\pi x$

K. A rectangular solid with dimensions x, $\frac{1}{3}x$, πx

51. A positive integer is called a palindrome if it reads the same forward as it does backward. For example, 1001 and 343 are palindromes. The next two palindromes greater than 9999 are m and p, where $m < p$. What is the value of $p - m$?

A. 10
B. 90
C. 100
D. 110
E. 210

52. Robert bought a book, and all of its pages are numbered in order. In the exact middle of the book, the product of two page numbers on the facing pages was 600. How many numbered pages are in the book?

F. 44
G. 46
H. 48
J. 50
K. 52

53. For nonzero numbers a and b, let $a \# b$ be defined by $a \# b = \frac{b^2}{a} + \frac{a^2}{b}$. If $c \neq 0$ and $c \# c = 4$, what is the value of c?

A. $\frac{1}{2}$

B. 2
C. 4
D. 6
E. 8

54. If $y^2 = \frac{2}{4 - 3x}$, then $x =$

F. $\frac{4}{3} - \frac{2}{3\sqrt{y}}$

G. $\frac{4}{3} - \frac{2y^2}{3}$

H. $\frac{4}{3} - \frac{2}{3y^2}$

J. $\frac{2}{3} - \frac{4y^2}{3}$

K. $\frac{2}{3} - \frac{4}{3y^2}$

GO ON TO THE NEXT PAGE.

55. If -1 and 3 are both zeros of the polynomial g(x), then a factor of g(x) is

A. $x^2 - 2x - 3$
B. $x^2 - 8$
C. $x^2 - 2x + 8$
D. $x^2 + 8$
E. $x^2 - 2x - 8$

56. If $f(x) = 4x + 4$ and $f^{-1}(g(1)) = 3$, which of the following could be $g(x)$?

F. $10x + 2$
G. $20x + 5$
H. $4x^2 + 12$
J. $7x^2$
K. $\frac{x}{7}$

57. What is the maximum value of the function $f(x) = -x^2 + 2x + 3$?

A. -4
B. 0
C. 2
D. 4
E. 15

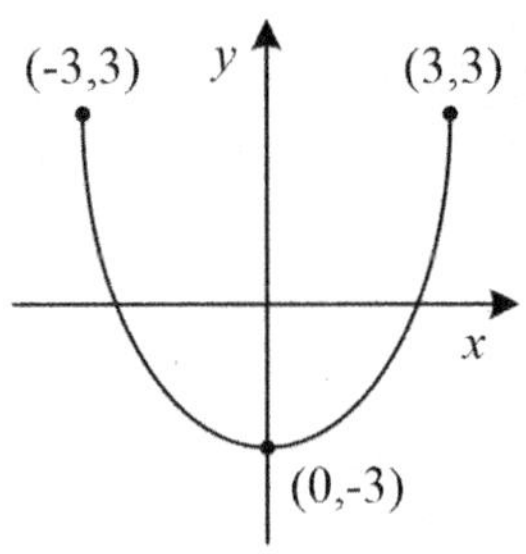

58. The graph of $y = f(x)$ is shown above. Which of the following statements must be true?

I. $f(x)$ is an even function.
II. The graph of $y = f(x)$ contains a vertical asymptote at $x = 0$.
III. The range and domain of $f(x)$ are equal.
IV. $f(-x) = -f(x)$

F. I only
G. II only
H. III only
J. I and III only
K. I, II, and IV

59. If $x = \log_2 t$ and $y = t^2$, what is y in terms of x?

A. 2
B. x
C. 2^x
D. 2^{2x}
E. 2^{x^2}

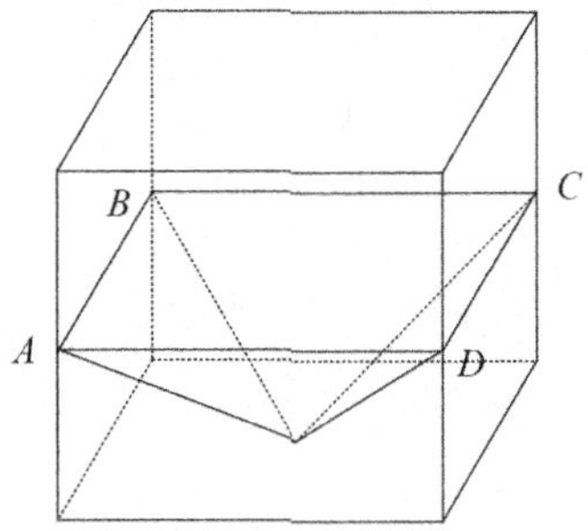

60. Above figure shows a pyramid inscribed in a cube. If points A, B, C, and D are midpoints of the edges of the cube, what is the ratio of the volume of the pyramid to the volume of the cube?

F. 1:3
G. 2:5
H. 3:5
J. 1:6
K. 2:9

END OF TEST 2
STOP! DO NOT TURN THE PAGE UNTIL TOLD TO DO SO.
DO NOT RETURN TO THE PREVIOUS TEST.

Answer Key
Model Test No. 02

1	C	16	G	31	E	46	F
2	K	17	E	32	G	47	C
3	E	18	F	33	C	48	H
4	H	19	C	34	J	49	A
5	A	20	K	35	B	50	F
6	F	21	E	36	F	51	C
7	D	22	F	37	C	52	H
8	H	23	A	38	J	53	B
9	C	24	F	39	B	54	H
10	F	25	A	40	F	55	A
11	E	26	J	41	B	56	H
12	H	27	C	42	F	57	D
13	C	28	H	43	C	58	J
14	H	29	B	44	K	59	D
15	A	30	J	45	C	60	J

How to Score the ACT Mathematics Test

When you take an actual ACT Mathematics Test, your answer sheet will be "read" by a scanning machine that will record your responses to each question. Then a computer will compare your answers with the correct answers and produce your raw score. You get one point for each correct answer. For each wrong answer, there is no penalty. Therefore, you must guess and answer all the questions to get a higher score. This raw score is converted to a scaled score that is reported to you and to the colleges you specify.

Finding Your Scale Test Score

STEP 1: Count the number of right answers.
Enter the total here: ______________________

STEP 2: The number you obtained in Step 1 is your raw score.
Look up the conversion table for ACT-Math, and find your score.

Scaled Score Conversion Table
ACT Mathematics Test

Scale Score	Mathematics
36	60
35	58-59
34	56-57
33	55
32	54
31	52-53
30	50-51
29	48-49
28	46-47
27	43-45
26	41-42
25	39-40
24	37-38
23	35-36
22	33-34
21	31-32
20	30
19	27-29
18	25-26
17	22-24
16	18-21
15	15-17
14	12-14
13	09-11
12	08
11	06-07
10	05
09	04
08	-
07	03
06	02
05	-
04	-
03	01
02	-
01	00

1. Auto Sales leases a new automobile for a down payment of \$3,200 plus \$480 per month for 36 months. Which of the following functions models the total amount, in dollars, at the end of m months of the lease?

   A. $f(m) = 3{,}200 + m$
   B. $f(m) = 3{,}200 + 36m$
   C. $f(m) = 3{,}200 + 480m$
   D. $f(m) = 3{,}200 + 480(36 - m)$
   E. $f(m) = 3{,}680 + 480(m - 1)$

The total amount paid at the end of m months with \$3,200 one-time down payment would be, $f(m) = 3{,}200 + 480m$.

Ans. C

2. If $y = x^2 - 2x - 1$ and x is a real number, which of the following CANNOT be the value of y?

   F. -2 **G.** -1 **H.** 0 **J.** 1 **K.** -5

$y = x^2 - 2x - 1 = (x-1)^2 - 2$, or $y + 2 = (x-1)^2 \geq 0$
$\therefore y \geq -2$, or vertex $v = (1, -2)$ and concave up!!
Therefore, y can be any of the values -2, -1, 0, 1.

Ans. K

x	$f(x)$
-1	1
-2	-2
-3	4
-5	6
-6	5

3. The function f is defined by the table above. For which value of x does $f(x) = 5$?

   A. -1 **B.** -2 **C.** -3 **D.** -5 **E.** -6

The only corresponding x value for $f(x) = 5$ is, when $x = -6$ from the given table.

Ans. E

4. If $x^2 + 4x \leq -3$, which of the following is a possible value of x?

   F. -6 **G.** -4 **H.** -2 **J.** 0 **K.** 2

$x^2 + 4x + 3 \leq 0$, or $(x+3)(x+1) \leq 0$.

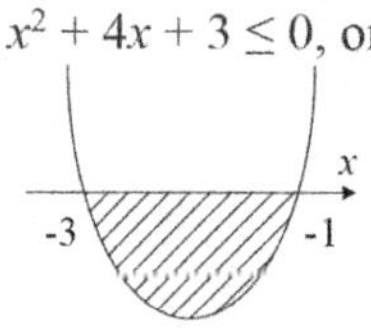

The figure shows, $-3 \leq x \leq -1$.

Ans. H

5. If $2n + p = 5$ and $2m - 4n = 2$, what is the value of $m + p$?

   A. 6 **B.** 4 **C.** 5 **D.** 7 **E.** 8

$2n + p = 5$
$\therefore 2n = 5 - p$, or $4n = 10 - 2p$
Now, substituting into the second equation, $2m - 4n = 2$,
we get $2m - (10 - 2p) = 2$, or $2m + 2p = 12$.
$\therefore m + p = 6$

Ans. A

6. If $3^y = z$, which of the following must be equal to 3^{y-1}?

   F. $\frac{z}{3}$ **G.** $z + 3$ **H.** $3z$ **J.** $\frac{z}{9}$ **K.** $\frac{z^2}{3}$

Here, we may use the properties of logarithmic function, but it is not necessary. Since $3^{y-1} = \frac{3^y}{3}$, we may replace $3^y = z$ into the above equation, we get $3^{y-1} = \frac{3^y}{3} = \frac{z}{3}$.

Ans. F

7. If n is a two-digit number and the sum of its units digit and tens digit is 9, which of the following statements must be true?

A. n is a multiple of 4.
B. n is a multiple of 5.
C. n is a multiple of 6.
D. n is a multiple of 9.
E. n is a multiple of 13.

To solve this problem, let's try with our answer choices.
A. n is a multiple of 4. For example, $n = 16$.
However, the sum of $1 + 6 = 7 \neq 9$ $\therefore$ No
B. $n = 25$, but $2 + 5 = 7 \neq 9$ $\therefore$ No
C. $n = 36$ $\therefore 3 + 6 = 9$,
but when $n = 12$, we get $1 + 2 = 3 \neq 9$ $\therefore$ No
D. $n = 18$ $\therefore 1 + 8 = 9$ $\therefore$ Yes
also $n = 27$ $\therefore 2 + 7 = 9$ $\therefore$ Yes
In fact, it is true for all multiple of 9.

Ans. D

8. 10 percent of 50 is 50 percent of what number?

F. 40 **G.** 20 **H.** 10 **J.** 4 **K.** 2

$(0.1)(50) = (0.5)x$ $\therefore x = 10$

Ans. H

$k^2 \geq k^3$ for all real numbers k

9. Which of the following values for k shows that the statement above is <u>false</u>?

A. -1 **B.** 0 **C.** $\frac{3}{2}$ **D.** 1 **E.** $\frac{1}{2}$

Since k is a real number, all of our answer choices satisfy the condition, except $k = \frac{3}{2}$. To hold $k^2 \geq k^3$ be true, we must have $0 \leq k \leq 1$. Let $k = \frac{3}{2}$, then $\left(\frac{3}{2}\right)^2 \leq \left(\frac{3}{2}\right)^3$.
$\therefore k = \frac{3}{2}$ does not hold for $k^2 \geq k^3$

Ans. C

10. If k is 10 percent of p and p is 20 percent of s, what percent of s is k?

F. 2% **G.** 4% **H.** 10% **J.** 40% **K.** 100%

$k = (0.1)p$, $p = (0.2)s$
The question is, $(x\ \%)(s) = k$ $\therefore \frac{x}{100} \cdot s = k$
Now, replacing $s = \frac{p}{0.2}$, and $k = (0.1)p$ into the above equation, we get, $\frac{x}{100} \cdot \frac{p}{0.2} = (0.1)p$ $\therefore \frac{x}{20} = 0.1$, or $x = 2$

Ans. F

11. A certain restaurant parlor offers 8 different toppings for pizzas. How many combinations of 3 different toppings from those offered are possible?

A. 6 **B.** 12 **C.** 15 **D.** 36 **E.** 56

We are choosing 3 toppings out of total 8 different toppings, regardless of certain orders. Therefore, it is a combination, not a permutation. $\therefore {}_8C_3 = 56$

Ans. E

$$\frac{4}{x-2} > 1$$

12. In which of the following intervals does every value of x satisfy the inequality above?

F. $-6 \leq x \leq -5$
G. $-5 \leq x \leq -3$
H. $3 \leq x \leq 5$
J. $2 \leq x \leq 6$
K. $6 \leq x \leq 8$

i. Let $x - 2 > 0$, then multiplying both side of inequality we get, $4 > (x - 2)$ $\therefore 6 > x$ Since $x - 2 > 0$, we get $2 < x < 6$.
ii. Now, let $x - 2 < 0$, then since $(x - 2)$ is negative, multiplying both side of inequality, we get $4 < (x - 2)$, or $x > 6$.
Since we assume $x - 2 < 0$, or $x < 2$, we have, $x < 2$ and $x > 6$, which DO NOT have solution!!

Therefore, only case **i.** works with $2 < x < 6$, which includes our answer choice **H**.
(Note: The answer choice **J** can't work, because $x \neq 2$ or $x \neq 6$)

Ans. H

x	y
1/6	3
1/3	6
5/9	?
2	36
3	54

13. In the table above, x and y are directly proportional. What is the value of y when x is $\frac{5}{9}$?

A. 8 **B.** 9 **C.** 10 **D.** 12 **E.** 18

$y = kx$

$\therefore 3 = k\frac{1}{6}$ $\therefore k = 18$ $\therefore y = 18x$

Now, when $x = \frac{5}{9}$, we get $y = 18\frac{5}{9} = 10$.

Ans. C

1, 2, 4, 8 …

14. The first term of the sequence above is 1, and each term after the first is twice the preceding term. Which of the following expressions represents the nth term of the sequence?

F. 2^n **G.** $2^n - 1$ **H.** 2^{n-1} **J.** $(n-1)^2$ **K.** $2(n-1)$

In geometric sequence, the general term, $a_n = a_1 r^{n-1}$.
Here, we have $a_1 = 1$, $r = 2$.
$\therefore a_n = 2^{n-1}$.

Ans. H

Alice's Monthly Budget

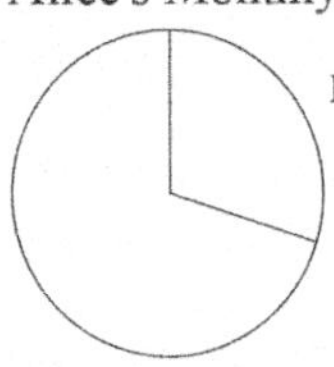

15. On the basis of the information in the graph above, if Alice's monthly housing budget is \$900, what is the dollar amount of her total monthly budget?

A. \$3,000 **B.** \$2,700 **C.** \$3,600 **D.** \$1,800 **E.** \$900

Let her total monthly budget is T, then (0.30)T = \$900

$\therefore T = \$\frac{900}{0.30} = \$3,000$

Ans. A

16. When 10^3 is added to 34,567, the value of which digit in the number 34,567 will be increased?

F. 3 **G.** 4 **H.** 5 **J.** 6 **K.** 7

$34,567 + 10^3 = 34,567 + 1,000 = 35,567$
Therefore, the digit 4 in the thousand digit increased to 5.

Ans. G

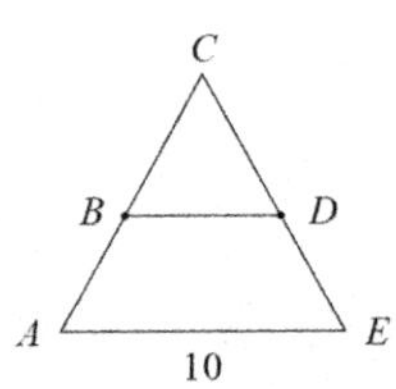

17. In ΔACE above, $AC = 2BC$ and $CE = 2CD$. If $AC = 9$ and $CE = 15$, what is the perimeter of ΔBCD?

A. 13 **B.** 14 **C.** 15 **D.** 16 **E.** 17

Let $x = BC$ and $y = CD$, then $AC = 9 = 2x$.
$\therefore x = 4.5$ and $CE = 15 = 2y$ $\therefore y = 7.5$.
Also, the points B and D are midpoint of $\overline{AC}$ and $\overline{CE}$ respectively, $\overline{BD}$ is $\frac{1}{2}$ of $\overline{AE}$. $\therefore \overline{BD} = \frac{1}{2} \times 10 = 5$
Therefore, the perimeter of ΔACE is,
$p = x + y + \overline{BD} = 4.5 + 7.5 + 5 = 17$

Ans. E

$N(p) = -1.5p + 45$

18. For a certain product, the number N of thousands of units sold each month is given by the function above, where p is the price per unit, in dollars, and $0 < p \leq 10$. The price per unit was \$10 in July and \$8 in August. How does the number of units sold in July compare to the number of units sold in August?

F. 3,000 fewer units were sold in July.
G. 8,000 fewer units were sold in July.
H. 20,000 fewer units were sold in July.
J. 12,000 more units were sold in July.
K. 16,000 more units were sold in July.

$N(10) = -1.5(10) + 45 = 30$
$N(8) = -1.5(8) + 45 = 33$
Therefore, $33 - 30 = 3$ thousand units.

Ans. F

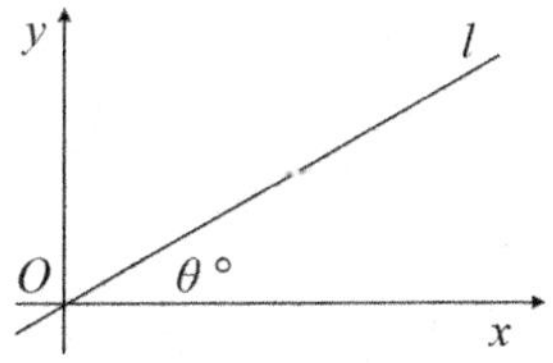

19. In the figure above, what is the equation of line l?

A. $y = x \sin \theta$
B. $y = x \cos \theta$
C. $y = x \tan \theta$
D. $y = x \csc \theta$
E. $y = x \sec \theta$

Since slope $m = \dfrac{(y_2 - y_1)}{(x_2 - x_1)} = \dfrac{\text{Rise}}{\text{Run}} = \dfrac{\text{Opposite}}{\text{Adjacent}} = \tan \theta$, we get $y = mx = (\tan \theta)\cdot x$.

Ans. C

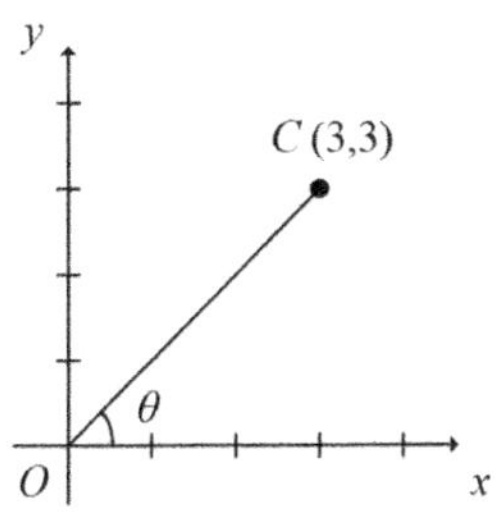

20. In the xy-coordinate system above, what is the angle θ formed by $\overline{OC}$ and x-axis?

F. 36° **G.** 18° **H.** 30° **J.** 60° **K.** 45°

Here, again $\tan \theta = \dfrac{y}{x} = \dfrac{3}{3} = 1$.
$\therefore \theta = \tan^{-1}(1) = 45°$

Ans. K

$|x - 5| < 3$
$|y + 2| < 7$

21. If x and y satisfy the inequalities above, then $|xy|$ could be any of the following EXCEPT

A. 18 **B.** 56 **C.** 64 **D.** 62 **E.** 80

$|x - 5| < 3$ becomes $-3 < x - 5 < 3$, or $2 < x < 8$.
And $|y + 2| < 7$ becomes $-7 < y + 2 < 7$, which is $-9 < y < 5$.
Therefore, we have $2 < x < 8$, and $-9 < y < 5$.
Now, from the above inequality, we have the max of $|xy| = |(-)9 \times 8| = 72$, and the min of $|xy| = |2 \times (0)| = 0$.
$\therefore 0 < xy < 72$
$\therefore$ 80 is excluded.

Ans. E

22. If the diagonals of quadrilateral P are perpendicular to each other, which of the following statements must be true?

I. All sides of P have equal length.
II. Each side of P is parallel to the opposite side.
III. The measure of each interior angle of P is 90°

F. None
G. I only
H. II only
J. III only
K. I, II, and III

When the diagonals of quadrilateral are perpendicular, it is either a kite or a rhombus as shown here.

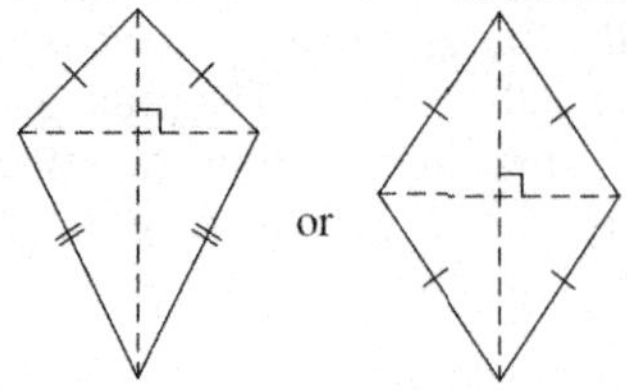

In case of a kite, it does not have to be either of (I), (II) and (III).

Ans. F

23. The first term of a sequence is 3 and the second term is 1. From the third term on, each term is the sum of all preceding terms. What is the 10th term of the sequence?

A. 512 **B.** 768 **C.** 1024 **D.** 1280 **E.** 2048

3, 1, (3 + 1), 3+ 1 + (3 + 1), … = 3, 1, 4, 8, 16, 32, 64, …
Here, we notice that from the 3rd term and on,
we get the next term being the twice of the previous term.
For example, the 4th term, $8 = 2^{4-1}$, the fifth term, $16 = 2^{5-1}$ …
$\therefore$ the tenth term becomes $2^{10-1} = 512$

Ans. A

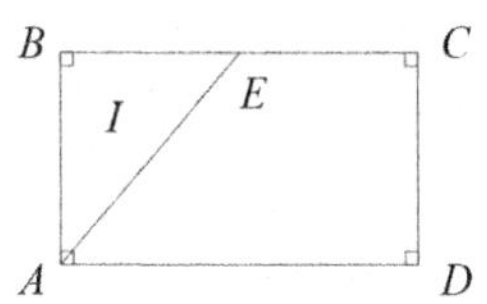

24. In the figure above, the area of triangle I is $\frac{1}{4}$ the area of rectangle $ABCD$, and the area of the triangle is 8. If the length of AB is equal to the length of BE, then what is the length of AD?

F. 8 **G.** 16 **H.** 12 **J.** 10 **K.** 18

Let $\overline{AB} = \overline{BE} = x$, then area of ΔABE is, $\frac{1}{2}x^2 = 8$
$\therefore x^2 = 16$
$\therefore x = 4$

Since the area of triangle I is $\frac{1}{4}$ of rectangle $ABCD$,
we may say E is the mid-point of $\overline{BC}$.
$\therefore \overline{BC} = \overline{AD} = 2\overline{BE} = 2 \times 4 = 8$

Ans. F

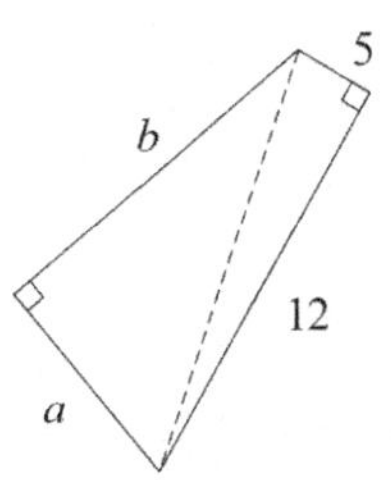

25. In the figure above, what is the value of $a^2 + b^2$?

A. 169 **B.** 144 **C.** 124 **D.** 256 **E.** 162

Using Pythagorean Theorem,
we get $\sqrt{a^2 + b^2} = \sqrt{5^2 + 12^2} = 13$
$\therefore a^2 + b^2 = 13^2 = 169$

Ans. A

26. In a race, Max and Tom started at the same time and ran with constant speeds of 10 kilometers and 16 kilometers per hour, respectively. If Max crossed the finish line 22.5 minute after Tom, how many kilometers long was the race?

F. 0.25 **G.** 0.5 **H.** 2.0 **J.** 10.0 **K.** 30.0

Let d = the distance of the race, and t = the time in hours that Tom finished. Then, using the fact that distance = speed × time, we get $d = 16 \times t$ - eq ①

$d = 10 \times (t + \frac{22.5}{60})$ - eq ②

Now, eq ① – eq ② = $16t - (10t + \frac{22.5}{6}) = 0$

$\therefore 6t = \frac{22.5}{6}$

$\therefore t = \frac{22.5}{36}$ hr.

Therefore, $d = 16 \times \frac{22.5}{36} = 10$.

Ans. J

27. If $x = t + 3$ and $y = 2t - 5$, which of the following is an expression for y, in terms of x?

A. $2x$ **B.** $2x + 11$ **C.** $2x - 11$ **D.** $2x - 1$ **E.** $2x + 1$

$x = t + 3$

$\therefore t = x - 3$

Now, replacing t into $y = 2t - 5$, we get $y = 2(x - 3) - 5 = 2x - 11$.

Ans. C

$$\begin{array}{r} \text{AD} \\ +\ \text{AV} \\ \hline \text{DVD} \end{array}$$

28. In the correctly worked addition problem above, A, D and V represent different digits. What is the value of A?

F. 0 **G.** 1 **H.** 5 **J.** 8 **K.** 9

Here, the sum of 2 digit numbers produce 3 digit number that is less than 199. Therefore, the hundred digit, D = 1.

Now, we get

$$\begin{array}{r} \text{A1} \\ +\ \text{AV} \\ \hline \text{1V1} \end{array}$$

Here, the sum of unit digit 1 + V = 11

∴ V = 0, and this leads us to conclude that A becomes 5.

∴ D = 1, V = 0, A = 5.

In fact,

$$\begin{array}{r} 51 \\ +50 \\ \hline 101 \end{array}$$

Ans. H

29. If $(7 \times 10^5) + (4 \times 10^n) = (7.4 \times 10^5)$, what is the value of n?

A. 3 **B.** 4 **C.** 6 **D.** 7 **E.** 9

Since $7.4 \times 10^5 = (7 + 0.4) \times 10^5$

$= 7 \times 10^5 + 0.4 \times 10^5$

$= 7 \times 10^5 + 4 \times 10^4$

$\therefore n = 4$

Ans. B

30. A large solid cube was assembled by gluing together small cubic blocks. All six faces of the large cube were then painted black. When the large cube was disassembled, 8 of the small cubic blocks had no black paint on their faces. How many small cubic blocks had black paint on any of their faces?

F. 36 **G.** 42 **H.** 48 **J.** 56 **K.** 64

8 of the unpainted small blocks consist a $2 \times 2 \times 2$ cube.

∴ The painted blocks = Total – inside blocks

$= (4 \times 4 \times 4) - (2 \times 2 \times 2) = 64 - 8 = 56$

Ans. J

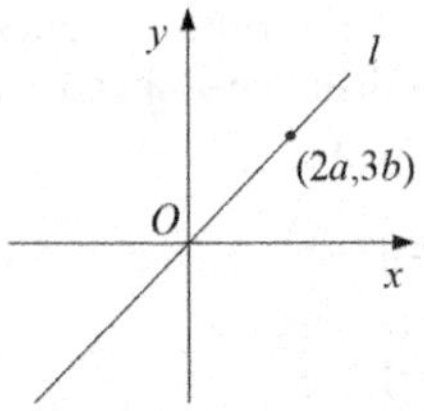

31. In the figure above, if $a = 4$ and $b = 2$, what is the slope of line l?

A. $\frac{3}{8}$ **B.** $\frac{4}{9}$ **C.** $\frac{1}{2}$ **D.** $\frac{5}{8}$ **E.** $\frac{3}{4}$

The slope $m = \frac{\text{rise}}{\text{run}} = \frac{3b}{2a} = \frac{3(2)}{2(4)} = \frac{3}{4}$

Ans. E

32. Points A and B are on a circle with center O. If the measure of angle AOB is 2 radian and $AO = 5$, what is the length of minor arc AB?

F. 9.78 **G.** 10 **H.** 10.34 **J.** 11 **K.** 11.16

Here, $\widehat{AB} = r \cdot \theta = 5 \cdot 2 = 10$

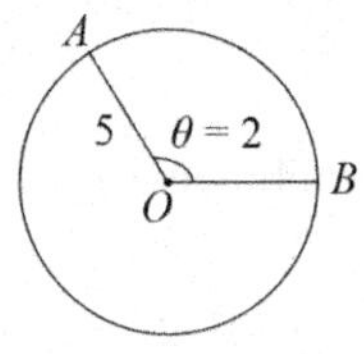

Ans. G

33. A circle with radius r has three non-overlapping arcs with lengths 3π, 5π, and 9π. If the radius is given by $r = 10$, then what is the remaining arc length that makes up the entire circle?

A. 2π
B. 4π
C. 3π
D. π
E. 5π

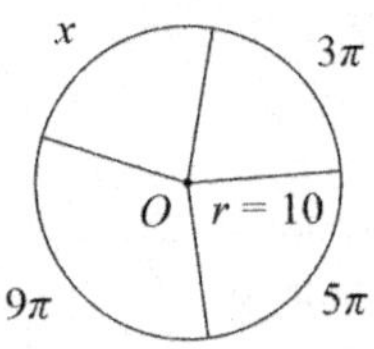

Since the circumference of the circle, $C = 2\pi r = 2\pi(10) = 20\pi$, we need the extra arc length of $x = 20\pi - 3\pi - 5\pi - 9\pi = 3\pi$

Ans. C

34. If 8 numbers are paired so that the four pairs of numbers have averages (arithmetic means) of 9, 13, 17 and 21, respectively, what is the sum of the 8 numbers?

F. 114 **G.** 115 **H.** 118 **J.** 120 **K.** 121

The sum, $S = (2 \times 9) + (2 \times 13) + (2 \times 17) + (2 \times 21) = 120$

Ans. J

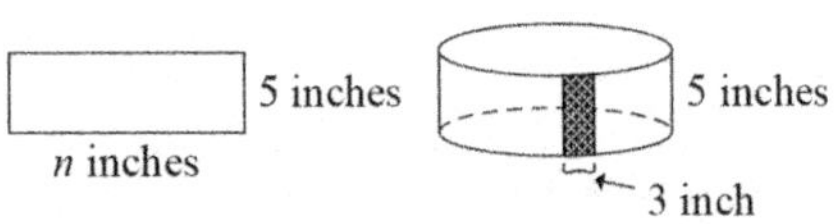

Figure I Figure II

35. The rectangular sheet of paper shown in Figure I above is rolled into the shape of a right circular cylinder, as shown in Figure II, The shaded region of width 3 inch shows where the two ends of the paper overlap. If the volume of the cylinder is $\frac{125}{\pi}$ cubic inches, what is the value of n?

A. 12 **B.** 13 **C.** 15 **D.** 19 **E.** 22

The volume of the cylinder,

$V = \pi r^2 h = \pi r^2(5) = \frac{125}{\pi}$

$\therefore r^2 = \frac{25}{\pi^2} = \left(\frac{5}{\pi}\right)^2 \quad \therefore r = \frac{5}{\pi}$

Now, the circumference, $C = 2\pi r = 2\pi\frac{5}{\pi} = 10$.

Since 3" is overlapped, we add 3.

$\therefore n = 10 + 3 = 13$

Ans. B

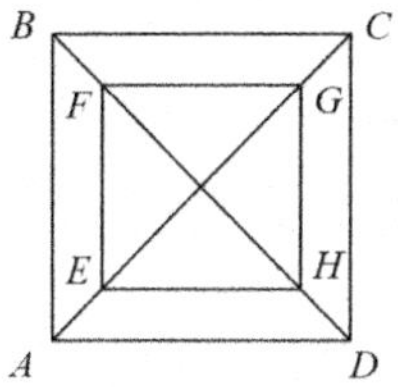

36. If the area of square $ABCD$ is 144, and $AE = BF = CG = DH = 3\sqrt{2}$, what is the perimeter of the square $EFGH$?

F. 24 **G.** 25 **H.** 28 **J.** 30 **K.** 31

Since the area of the square $ABCD$ is 144, we know that $\overline{AB} = \overline{BC} = \overline{CD} = \overline{AD} = 12$. We also know that the diagonal $\overline{AC} = 12\sqrt{2}$, because the figure is a square.
$\therefore \overline{EG} = \overline{AC} - \overline{AE} - \overline{GC} = 12\sqrt{2} - 3\sqrt{2} - 3\sqrt{2} = 6\sqrt{2}$
Again, since $\square EFGH$ is a square, and its diagonal $\overline{EG} = 6\sqrt{2}$, we get $\overline{EF} = \overline{FG} = \overline{GH} = \overline{HE} = 6$
$\therefore$ perimeter, $p = 24$

Ans. F

37. Jim cut a circular pie into wedge-shaped pieces. The tip of each piece at the center of the pie and the angle at the tip is always greater than 25°, but less than 30°. What is the least value for the number of pieces into which the pie is cut?

A. 10 **B.** 12 **C.** 13 **D.** 15 **E.** 16

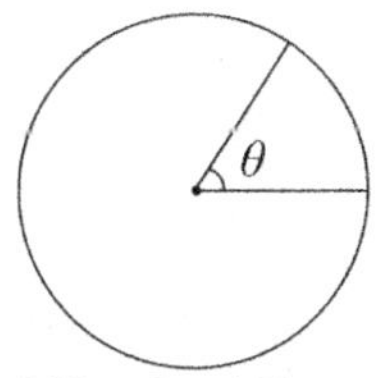

$25° < \theta < 30°$
$\therefore$ The number of the pie pieces, n would be between
$\frac{360°}{30°} < n < \frac{360°}{25°}$, or $12 < n < 14.4$
$\therefore n = 13$ will be the least one.

Ans. C

38. If $(x + 5)(x - 5) = 75$, what is the value of x^2?

F. 88 **G.** 94 **H.** 98 **J.** 100 **K.** 104

$(x + 5)(x - 5) = x^2 - 25 = 75$
$\therefore x^2 = 100$

Ans. J

39. On a bus trip to Seattle, all of the seats were occupied and 3 people were standing. On the return trip, 11 of the seats were empty and no one was standing. If the total bus fares were \$255 and \$185, respectively, for the two trips and all passengers paid the same fare, what was the passenger seating capacity of the bus?
(Assume one passenger per seat.)

A. 45 **B.** 48 **C.** 53 **D.** 55 **E.** 60

Let x = # of seats in one bus and m = the bus fare, then we get
$\$255 = m(x + 3)$, eq ①
$\$185 = m(x - 11)$, eq ②
Now, eq ① – eq ② gives, $70 = 3m + 11m = 14m$
$\therefore m = \$5.00$ for the fare.
Substituting back into eq(1), we get, $255 = 5(x + 3)$
$\therefore x = 48$

Ans. B

40. James has to sell a box of candy bars for a fund-raiser. Yesterday he sold $\frac{2}{5}$ of the candy bars in the box, and today he sold 16 candy bars. If his sales for these two days was $\frac{2}{3}$ of the total number of candy bars that came in the box, how many candy bars are left in the box?

F. 20 **G.** 30 **H.** 40 **J.** 60 **K.** 90

Let T = # of candy bars in the box. Then, $\frac{2}{5}T + 16 = \frac{2}{3}T$.
$\therefore 16 = \frac{2}{3}T - \frac{2}{5}T = \frac{4}{15}T \quad \therefore T = \frac{16 \times 15}{4} = 60$
But the number of candy bars sold is, $\frac{2}{3}T = \frac{2}{3}(60) = 40$.
$\therefore$ The number of the candy bars left becomes, $60 - 40 = 20$.

Ans. F

41. If k is a positive integer, then $(-1)^k(-1)^{k+1} =$

A. -2 **B.** -1 **C.** 1 **D.** $-2k - 1$ **E.** $\dfrac{1}{2k+1}$

$(-1)^k(-1)^{k+1} = (-1)^{2k+1}$, but $(2k + 1)$ = odd number
$\therefore (-1)^{2k+1} = -1$

Ans. B

42. The balls in a certain basket are either red or blue. If the ratio of red balls to blue balls is $\frac{1}{5}$, each of the following could be the number of balls in the basket EXCEPT

F. 15 **G.** 30 **H.** 42 **J.** 60 **K.** 90

Let R = # of red balls, and B = # of blue balls. Then, $\frac{R}{B} = \frac{1}{5}$.
$\therefore$ B = 5R, and the total, T = B + R = 5R + R = 6R.
Since T is a multiple of 6's, the only choice that is <u>not</u> a multiple of 6's would be **F.** 15.

Ans. F

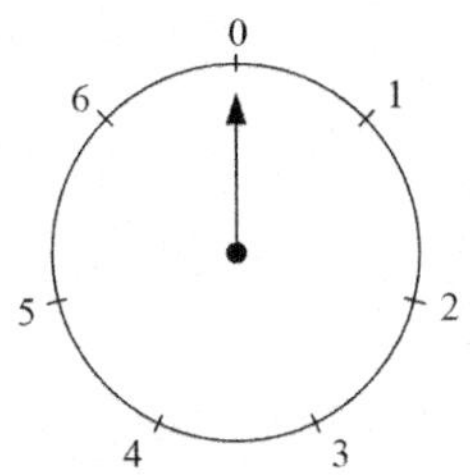

43. On the dial shown above, the numbers are equally spaced. The arrow rotates clockwise at a constant rate through an angle of 100° every 10 seconds. If the arrow starts at 0, how many times will the arrow pass the number 2 on the dial in 2 minutes?

A. Two **B.** Three **C.** Four **D.** Five **E.** Six

Since 2 min = 2 × 60 sec = 120 sec,
and also the angular speed is, $w = \frac{100°}{10 \text{ sec}} = \frac{10°}{\text{sec}}$,
we get $\frac{10°}{\text{sec}} \times 120 \text{ sec} = 1200°$.
Since one rotation is 360°, 1200° ÷ 360° = 3 rotation plus 120°.
But the number 2 is exactly on 120°.
$\therefore$ 3 + 1 = 4 times.

Ans. C

44. If x is a positive integer and if $y = x + \frac{1}{x}$, which of the following must be true?

I. $y \geq ?$
II. $xy \geq 1$
III. $xy > x^2$

F. I only
G. III only
H. I and II only
J. I and III only
K. I, II, and III

Let $x = 1$, then $y = 1 + \frac{1}{1} = 2$.
Then, (I) $y \geq 2$ is true.
(II) $xy \geq 1$ is true.
(III) $xy \geq x^2$ is true.

Ans. K

45. Points A, B, and C are on line l in that order. P is a point not on l. The measure of $\angle ABP$ is $(6x - 15)°$ and the measure of $\angle PBC$ is $(2x + 16)°$. What is the value of x?

A. 11.75 **B.** 21.5 **C.** 22.4 **D.** 23 **E.** 23.5

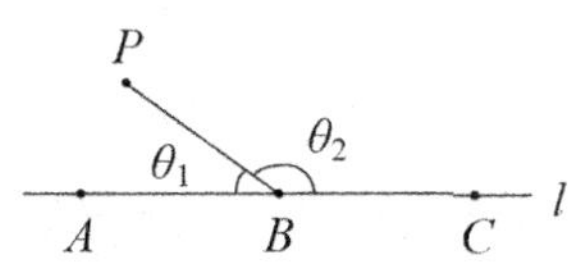

Here, $\theta_1 = 6x - 15$ and $\theta_2 = 2x + 16$, and also $\theta_1 + \theta_2 = 180°$.
$\therefore (6x - 15) + (2x + 16) = 8x + 1 = 180°$
$\therefore 8x = 179°$ $\therefore x = 22.4°$

Ans. C

46. Which of the following is one possible value of k for which $k < 3 < \frac{1}{k}$?

F. $\frac{1}{4}$ **G.** $\frac{1}{2}$ **H.** $\frac{1}{3}$ **J.** 2 **K.** 4

Let's just plug in our answer choices!

Then, if $k = \frac{1}{4}$, we get $k < 3 < \frac{1}{k}$ becomes $\frac{1}{4} < 3 < \frac{1}{\frac{1}{4}}$ or $\frac{1}{4} < 3 < 4$, which works.

Ans. F

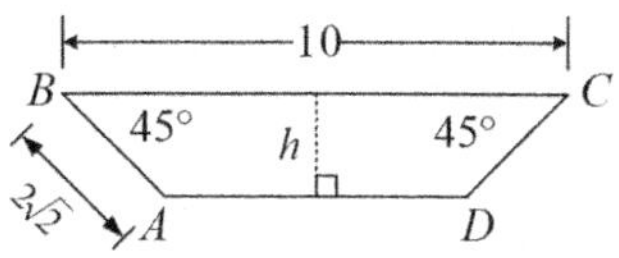

47. In the figure above, if $\overline{BC} \parallel \overline{AD}$ and $\angle ABC = \angle DCB$, what is the area of $ABCD$?

A. 14 **B.** 15 **C.** 16 **D.** 17 **E.** 18

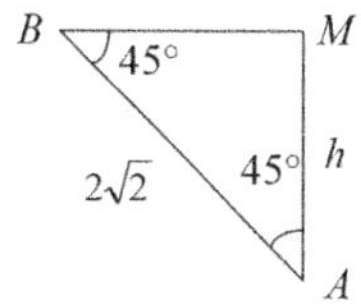

Considering a corner triangle, we get $h = 2$, since it is 45° - 45° - 90°.

$\therefore \overline{BM} = 2$

$\therefore \overline{AD} = 10 - 2 - 2 = 6$

$\therefore$ The area $= \frac{1}{2}(10 + 6) \cdot h = \frac{1}{2}(10 + 6) \cdot 2 = 16$

Ans. C

48. Brad and Patrick can each mow the lawn in an hour, and their father can mow the lawn in 30 minutes. If all three work together at their respective rates, how many minutes will it take to mow the lawn?

F. 5 **G.** 10 **H.** 15 **J.** 25 **K.** 30

Let B = Speed of Brad, P = Speed of Patrick, and F = Speed of father, and W = 1 unit of work.
Then, the total speed of all three working together would be,

$$S = B + P + F = \frac{1}{60} + \frac{1}{60} + \frac{1}{30} = \frac{4}{60}$$

$$\therefore \text{Time} = \frac{\text{Work}}{\text{Speed}} = \frac{1}{\frac{4}{60}} = \frac{60}{4} = 15 \text{ min.}$$

Ans. H

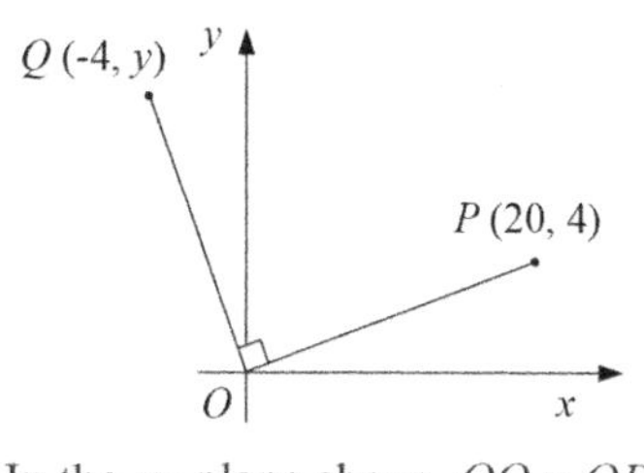

49. In the xy-plane above, $OQ = OP$. What is the value of y?

A. 20 **B.** 10 **C.** 14 **D.** 12 **E.** 8

Since $OQ \perp OP$,
the product of the slopes of OQ and OP becomes (-1).

$$\therefore \frac{4}{20} \times \frac{y}{-4} = -1$$

$$\therefore y = 20$$

Ans. A

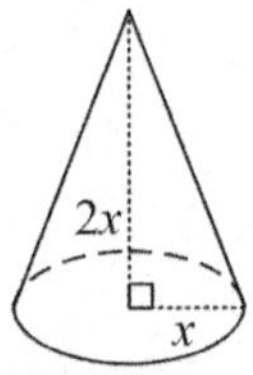

50. Which of the following has the same volume as the cone shown above with radius x and height $2x$?

F. A circular cylinder with radius $\frac{x}{3}$ and height $6x$

G. A circular cylinder with radius $\frac{2\pi x}{3}$ and height $3x$

H. A cube with edge $2x$

J. A cube with edge $2\pi x$

K. A rectangular solid with dimensions $x, \frac{1}{3}x, \pi x$

$V_1 = \frac{1}{3}\cdot\pi r^2 h = \frac{1}{3}\cdot\pi(x)^2(2x) = \frac{2}{3}\pi x^3$ for the cone.

Now, try answer choice (A). The volume of cylinder,

$V_2 = \pi R^2 H = \pi\left(\frac{x}{3}\right)^2\cdot(6x) = \frac{6}{9}\pi x^3 = \frac{2}{3}\pi x^3$,

which is the same volume as V_1.

Ans. F

51. A positive integer is called a palindrome if it reads the same forward as it does backward. For example, 1001 and 343 are palindromes. The next two palindromes greater than 9999 are m and p, where $m < p$. What is the value of $p - m$?

A. 10 **B.** 90 **C.** 100 **D.** 110 **E.** 210

Here, $m = 10001$ and $p = 10101$.

$\therefore p - m = 10101 - 10001 = 100$

Ans. C

52. Robert bought a book, and all of its pages are numbered in order. In the exact middle of the book, the product of two page numbers on the facing pages was 600. How many numbered pages are in the book?

F. 44 **G.** 46 **H.** 48 **J.** 50 **K.** 52

Let the front page number $= x$. Then $x(x + 1) = 600$.

$\therefore x^2 + x - 600 = 0$, or $(x + 25)(x - 24) = 0$

$\therefore x = 24$ and $x \neq 25$ since x is a page number.

$\therefore 24 \times 2 = 48$

Ans. H

53. For nonzero numbers a and b, let $a \# b$ be defined by $a \# b = \frac{b^2}{a} + \frac{a^2}{b}$. If $c \neq 0$ and $c \# c = 4$, what is the value of c?

A. $\frac{1}{2}$ **B.** 2 **C.** 4 **D.** 6 **E.** 8

According to the definition of $a \# b$,

$c \# c = \frac{c^2}{c} + \frac{c^2}{c} = c + c = 2c = 4$

$\therefore c = 2$

Ans. B

54. If $y^2 = \frac{2}{4 - 3x}$, then $x =$

F. $\frac{4}{3} - \frac{2}{3\sqrt{y}}$ **G.** $\frac{4}{3} - \frac{2y^2}{3}$ **H.** $\frac{4}{3} - \frac{2}{3y^2}$ **J.** $\frac{2}{3} - \frac{4y^2}{3}$ **K.** $\frac{2}{3} - \frac{4}{3y^2}$

The reciprocal of the equation would be, $\frac{1}{y^2} = \frac{4 - 3x}{2}$.

$\therefore$ multiplying 2 on both sides,

we get $\frac{2}{y^2} = 4 - 3x$, or $3x = 4 - \frac{2}{y^2}$. $\therefore x = \frac{4}{3} - \frac{2}{3y^2}$

Ans. H

55. If -1 and 3 are both zeros of the polynomial g(x), then a factor of g(x) is

A. $x^2 - 2x - 3$ **B.** $x^2 - 8$ **C.** $x^2 - 2x + 8$
D. $x^2 + 8$ **E.** $x^2 - 2x - 8$

According to Factor Theorem, if $x = a$ is a zero of polynomial function $p(x)$, then $p(a) = 0$.

Therefore, if -1 and 3 are zeros of $g(x)$, then there must be $(x + 1)$ $(x - 3)$ factors.

Thus $(x + 1)(x - 3) = x^2 - 2x - 3$ is also a factor of $g(x)$.

Ans. A

56. If $f(x) = 4x + 4$ and $f^{-1}(g(1)) = 3$, which of the following could be $g(x)$?

F. $10x + 2$
G. $20x + 5$
H. $4x^2 + 12$
J. $7x^2$
K. $\frac{x}{7}$

First, we have $f^{-1}(g(1)) = 3$, which is inverse function of $f(x)$.
By applying "f" function on both sides,
we get $f[f^{-1}(g(1))] = f(3)$.
Since $f(f^{-1})$ = identity which cancels each other,
we have $g(1) = f(3)$.
But $f(3) = 4(3) + 4 = 16$,
which equals $g(1)$, we should find $g(1) = 16$.
The answer **H** gives $g(1) = 4(1)^2 + 12 = 16$.

Ans. H

57. What is the maximum value of the function $f(x) = -x^2 + 2x + 3$?

A. -4
B. 0
C. 2
D. 4
E. 15

Using the vertex form, $f(x) = -(x^2 - 2x) + 3 = -(x - 1)^2 + 4$. Therefore, we have maximum, $y = 4$ when $x = 1$, where the vertex is, $V = (1, 4)$.

Or using the formula, the maximum happens at $x = -\frac{b}{2a}$, where $a = -1$, $b = 2$, $c = 3$.

$\therefore x = \frac{-2}{-2} = 1$

$\therefore f(1) = -1 + 2 + 3 = 4$, the maximum

Ans. D

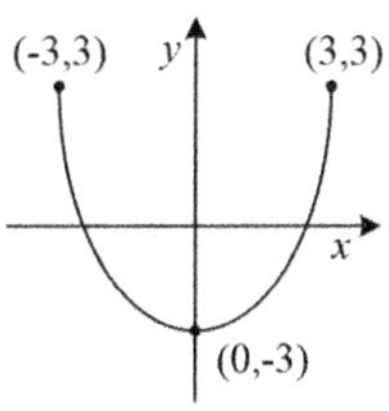

58. The graph of $y = f(x)$ is shown above.
Which of the following statements must be true?

I. $f(x)$ is an even function.
II. The graph of $y = f(x)$ contains a vertical asymptote at $x = 0$.
III. The range and domain of $f(x)$ are equal.
IV. $f(-x) = -f(x)$

F. I only
G. II only
H. III only
J. I and III only
K. I, II, and IV

I. Yes, $f(x)$ is symmetric to y-axis.
$\therefore$ Even function
II. There is no asymptotes
III. Yes, the range [-3,3] is equal to domain [-3,3].
IV. It should be $f(-x) = f(x)$ for even function.
$\therefore$ I and III only

Ans. J

59. If $x = \log_2 t$ and $y = t^2$, what is y in terms of x?

A. 2
B. x
C. 2^x
D. 2^{2x}
E. 2^{x^2}

By the definition of logarithmic function, if $a^y = x$, then $y = \log_a x$, we have $t = 2^x$ from $x = \log_2 t$.
Therefore, $y = t^2 = (2^x)^2 = 2^{2x}$

Ans. D

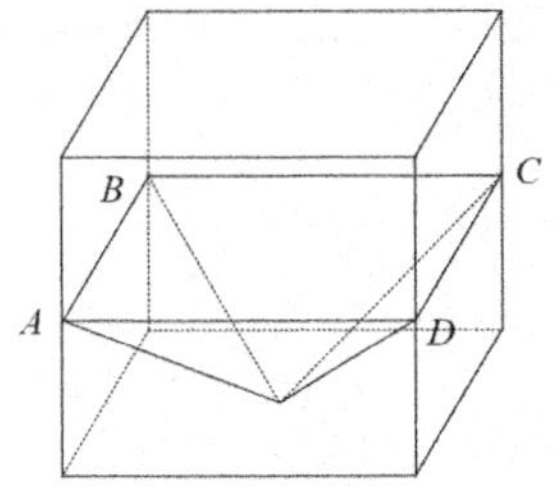

60. Above figure shows a pyramid inscribed in a cube. If points A, B, C, and D are midpoints of the edges of the cube, what is the ratio of the volume of the pyramid to the volume of the cube?

F. 1:3 **G.** 2:5 **H.** 3:5 **J.** 1:6 **K.** 2:9

Let the edge of each side of the cube be 2.
Then, the volume of the cube is $2 \times 2 \times 2 = 8$.
Since A, B, C, and D are all midpoints, the half of the cube has its volume equal to 4. But any shape of the pyramid is one-third of the volume that contains it. Therefore, the volume of the pyramid is $\frac{1}{3} \times 4 = \frac{4}{3}$. So, the ratio of the volume between the cube and pyramid is $\frac{4}{3} : 8 = 1:6$.

Ans. J

Model Test No. 03

60 Questions / 60 Minutes

Directions: Solve each problem, choose the correct answer, and then fill in the corresponding oval on your answer document.

Do not linger over problems that take too much time. Solve as many as you can; then return to the others in the time you have left for this test.

You are permitted to use a calculator on this test. You many use your calculator for any problems you choose, but some of the problems may best be done without using a calculator.

Note: Unless otherwise stated, all of the following should be assumed.

1. Illustrative figures are NOT necessarily drawn to scale.
2. Geometric figures lie in a plane.
3. The word line indicates a straight line.
4. The word average indicates arithmetic mean.

Answer Sheet
Model Test No. 03

1	Ⓐ Ⓑ Ⓒ Ⓓ Ⓔ	16	Ⓕ Ⓖ Ⓗ Ⓙ Ⓚ	31	Ⓐ Ⓑ Ⓒ Ⓓ Ⓔ	46	Ⓕ Ⓖ Ⓗ Ⓙ Ⓚ
2	Ⓕ Ⓖ Ⓗ Ⓙ Ⓚ	17	Ⓐ Ⓑ Ⓒ Ⓓ Ⓔ	32	Ⓕ Ⓖ Ⓗ Ⓙ Ⓚ	47	Ⓐ Ⓑ Ⓒ Ⓓ Ⓔ
3	Ⓐ Ⓑ Ⓒ Ⓓ Ⓔ	18	Ⓕ Ⓖ Ⓗ Ⓙ Ⓚ	33	Ⓐ Ⓑ Ⓒ Ⓓ Ⓔ	48	Ⓕ Ⓖ Ⓗ Ⓙ Ⓚ
4	Ⓕ Ⓖ Ⓗ Ⓙ Ⓚ	19	Ⓐ Ⓑ Ⓒ Ⓓ Ⓔ	34	Ⓕ Ⓖ Ⓗ Ⓙ Ⓚ	49	Ⓐ Ⓑ Ⓒ Ⓓ Ⓔ
5	Ⓐ Ⓑ Ⓒ Ⓓ Ⓔ	20	Ⓕ Ⓖ Ⓗ Ⓙ Ⓚ	35	Ⓐ Ⓑ Ⓒ Ⓓ Ⓔ	50	Ⓕ Ⓖ Ⓗ Ⓙ Ⓚ
6	Ⓕ Ⓖ Ⓗ Ⓙ Ⓚ	21	Ⓐ Ⓑ Ⓒ Ⓓ Ⓔ	36	Ⓕ Ⓖ Ⓗ Ⓙ Ⓚ	51	Ⓐ Ⓑ Ⓒ Ⓓ Ⓔ
7	Ⓐ Ⓑ Ⓒ Ⓓ Ⓔ	22	Ⓕ Ⓖ Ⓗ Ⓙ Ⓚ	37	Ⓐ Ⓑ Ⓒ Ⓓ Ⓔ	52	Ⓕ Ⓖ Ⓗ Ⓙ Ⓚ
8	Ⓕ Ⓖ Ⓗ Ⓙ Ⓚ	23	Ⓐ Ⓑ Ⓒ Ⓓ Ⓔ	38	Ⓕ Ⓖ Ⓗ Ⓙ Ⓚ	53	Ⓐ Ⓑ Ⓒ Ⓓ Ⓔ
9	Ⓐ Ⓑ Ⓒ Ⓓ Ⓔ	24	Ⓕ Ⓖ Ⓗ Ⓙ Ⓚ	39	Ⓐ Ⓑ Ⓒ Ⓓ Ⓔ	54	Ⓕ Ⓖ Ⓗ Ⓙ Ⓚ
10	Ⓕ Ⓖ Ⓗ Ⓙ Ⓚ	25	Ⓐ Ⓑ Ⓒ Ⓓ Ⓔ	40	Ⓕ Ⓖ Ⓗ Ⓙ Ⓚ	55	Ⓐ Ⓑ Ⓒ Ⓓ Ⓔ
11	Ⓐ Ⓑ Ⓒ Ⓓ Ⓔ	26	Ⓕ Ⓖ Ⓗ Ⓙ Ⓚ	41	Ⓐ Ⓑ Ⓒ Ⓓ Ⓔ	56	Ⓕ Ⓖ Ⓗ Ⓙ Ⓚ
12	Ⓕ Ⓖ Ⓗ Ⓙ Ⓚ	27	Ⓐ Ⓑ Ⓒ Ⓓ Ⓔ	42	Ⓕ Ⓖ Ⓗ Ⓙ Ⓚ	57	Ⓐ Ⓑ Ⓒ Ⓓ Ⓔ
13	Ⓐ Ⓑ Ⓒ Ⓓ Ⓔ	28	Ⓕ Ⓖ Ⓗ Ⓙ Ⓚ	43	Ⓐ Ⓑ Ⓒ Ⓓ Ⓔ	58	Ⓕ Ⓖ Ⓗ Ⓙ Ⓚ
14	Ⓕ Ⓖ Ⓗ Ⓙ Ⓚ	29	Ⓐ Ⓑ Ⓒ Ⓓ Ⓔ	44	Ⓕ Ⓖ Ⓗ Ⓙ Ⓚ	59	Ⓐ Ⓑ Ⓒ Ⓓ Ⓔ
15	Ⓐ Ⓑ Ⓒ Ⓓ Ⓔ	30	Ⓕ Ⓖ Ⓗ Ⓙ Ⓚ	45	Ⓐ Ⓑ Ⓒ Ⓓ Ⓔ	60	Ⓕ Ⓖ Ⓗ Ⓙ Ⓚ

MATHEMATICS TEST

60 Minutes – 60 Questions

DIRECTIONS: Solve each problem, choose the correct answer, and then fill in the corresponding oval on your answer document.

Do not linger over problems that take too much time. Solve as many as you can; then return to the others in the time you have left for this test.

You are permitted to use a calculator on this test. You many use your calculator for any problems you choose, but some of the problems may best be done without using a calculator.

Note: Unless otherwise stated, all of the following should be assumed.

1. Illustrative figures are NOT necessarily drawn to scale.
2. Geometric figures lie in a plane.
3. The word line indicates a straight line.
4. The word average indicates arithmetic mean.

---

1. If the average (arithmetic mean) of 2, 5, and x is 7, what is the average of 5, 8, and x?

A. 5
B. 6
C. 7
D. 8
E. 9

2. If $x + y = 2z$ and $z = 4$, what is the value of $x + y + z$?

F. 8
G. 12
H. 16
J. 24
K. 64

STUDENTS IN THREE MATH SECTIONS

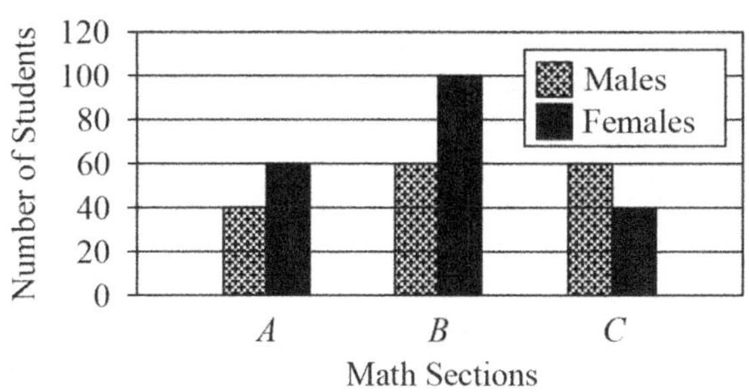

3. The graph above shows the numbers of students in 3 different Math Sections A, B and C. If these sections are to be combined into one section, what will be the ratio of the number of males to the number of females in the new section?

A. 2:3
B. 3:2
C. 3:5
D. 4:5
E. 5:4

4. Machine A produces a certain auto part at a constant rate of 5 per hour, and machine B produces at a constant rate of 8 per hour. How many more minutes does it take machine A to produce 20 auto parts than it takes machine B?

F. 20
G. 60
H. 80
J. 90
K. 120

$$\frac{1}{(1)(2)}, \frac{1}{(2)(3)}, \frac{1}{(3)(4)}$$

5. The n th term of a sequence is $\frac{1}{(n)(n+1)}$. What is the sum of the first 30 terms of this sequence?

A. 1
B. $\frac{30}{31}$
C. $\frac{29}{30}$
D. $\frac{29}{31}$
E. $\frac{14}{30}$

6. Which of the following is an equation of the line in the xy-plane that passes through the point (3, -3) and is perpendicular to the line $y = \frac{1}{2}x - 4$?

F. $y = \frac{1}{2}x + 3$
G. $y = -\frac{1}{2}x - 3$
H. $y = 2x - 3$
J. $y = -2x + 3$
K. $y = -2x - 3$

GO ON TO THE NEXT PAGE.

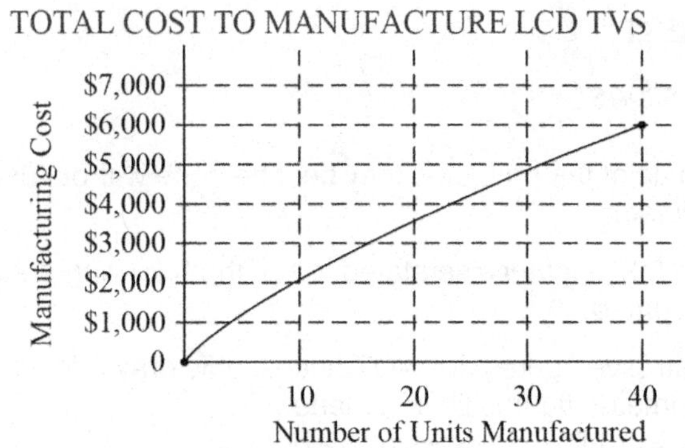

7. The graph above shows the total cost to manufacture a certain type of LCD TV based on the number of units manufactured. What is the approximate cost to manufacture 25 units?

A. \$3,500
B. \$4,400
C. \$5,200
D. \$5,700
E. \$6,100

8. Stained glasses measuring $\frac{1}{2}$ foot by $\frac{1}{3}$ foot are sold in boxes containing 10 each. What is the least number of boxes needed to cover a square window that has dimensions 12 feet by 12 feet?

F. 7
G. 16
H. 32
J. 57
K. 87

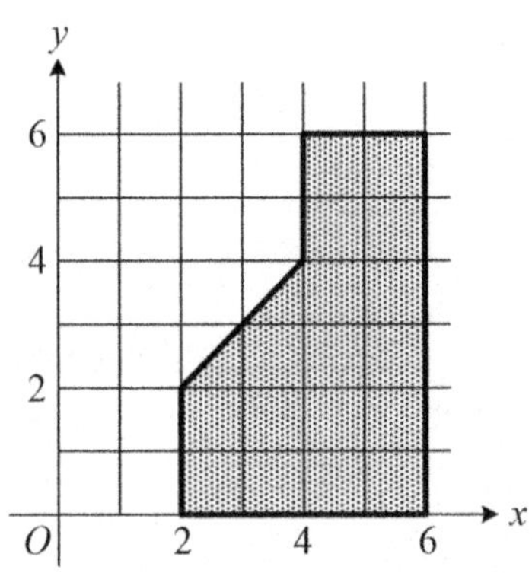

9. What is the area of the shaded part in the figure above?

A. 13
B. 14
C. 15
D. 18
E. 22

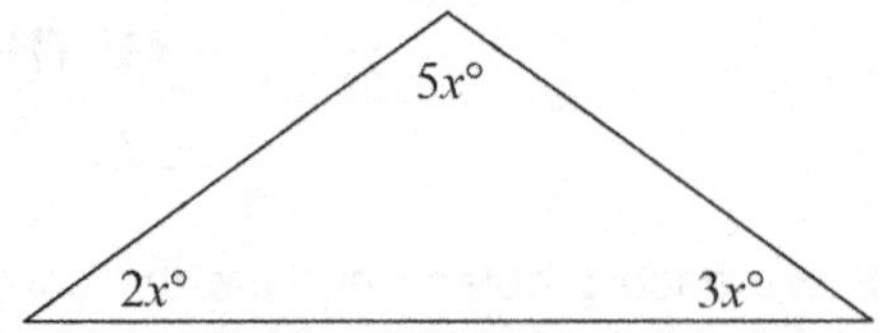

10. In the triangle above, what is the value of x?

F. 20
G. 18
H. 25
J. 15
K. 10

11. For expected high temperatures between 80 °F and 100 °F, the number of people in a water park on a given day is modeled by $5(x - 80)(100 - x)$, where x °F is the expected high temperature for that day. The expected high temperatures on July 4th and July 5th are 85 °F and 90 °F, respectively. According to this model, how many more people will be in the water park on July 5th than on July 4th?

A. 5
B. 125
C. 800
D. 900
E. 24,700

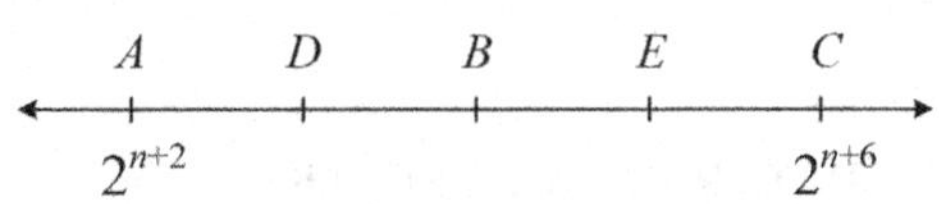

12. On the number line above, point B is the midpoint of $\overline{AC}$ and points D and E are the midpoint of $\overline{AB}$ and $\overline{BC}$ respectively. If n is positive, what is the value at point D?

F. $34 \cdot 2^n$
G. $14 \cdot 2^n$
H. $19 \cdot 2^n$
J. $36 \cdot 2^n$
K. $64 \cdot 2^n$

13. In a rectangular solid, eight of the edges each have a length of 8 inches and the other four each have a length of 9 inches. What is the volume, in cubic inches, of the rectangular solid?

A. 228
B. 576
C. 864
D. 1,728
E. 3,456

GO ON TO THE NEXT PAGE.

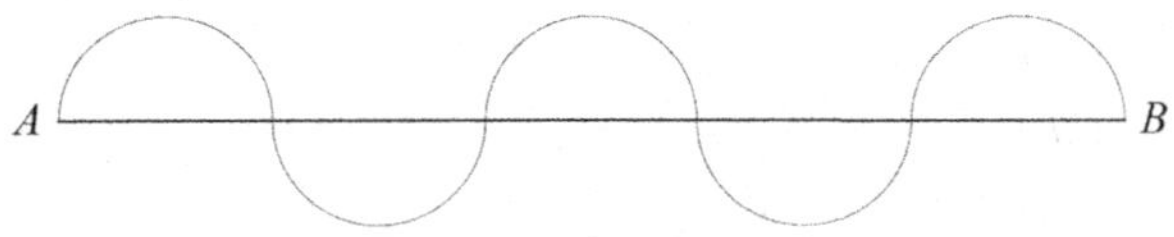

14. In the figure above, the diameters of the 5 semicircles are equal and lie on line segment $\overline{AB}$. If the length of line segment $\overline{AB}$ is 50, what is the area of the 5 semicircles from A to B?

F. 34π
G. 48π
H. 40π
J. 62.5π
K. 36.5π

15. If 5 UK quarts are approximately equal to 6 US quarts, which of the following is closest to 1 US quart?

A. 0.5 UK quart
B. 0.7 UK quart
C. 0.8 UK quart
D. 1.2 UK quart
E. 1.8 UK quart

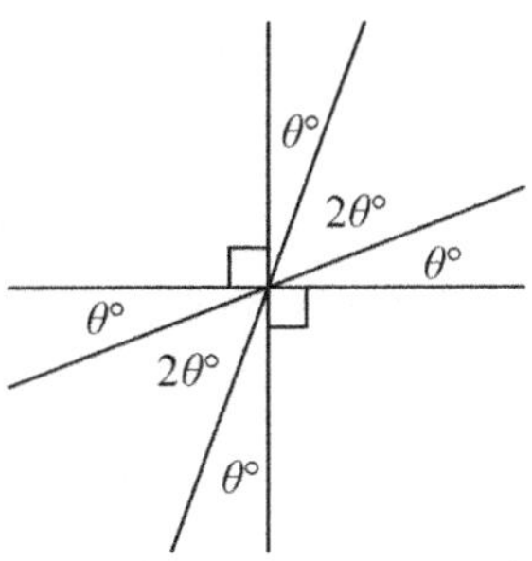

16. In the figure above, what is the value of θ?

F. $\frac{\pi}{4}$
G. $\frac{\pi}{6}$
H. $\frac{\pi}{8}$
J. $\frac{\pi}{10}$
K. $\frac{\pi}{12}$

17. If each class period is 45 minutes long, how many class periods are needed to show a 3-hour long video?

A. 1
B. 2
C. 3
D. 4
E. 5

18. The volume of a right circular cone is 18π cubic inches. If the height is twice of the radius of the cone, what is the height of the cone?

F. 4 inches
G. 6 inches
H. 8 inches
J. 16 inches
K. 32 inches

19. If $a > 0$, $x^2 + y^2 = a + 5$ and $xy = a - 5$, what is $(x - y)^2$ in terms of a?

A. $-a - 5$
B. $a - 5$
C. $a + 5$
D. $-a + 15$
E. $a - 15$

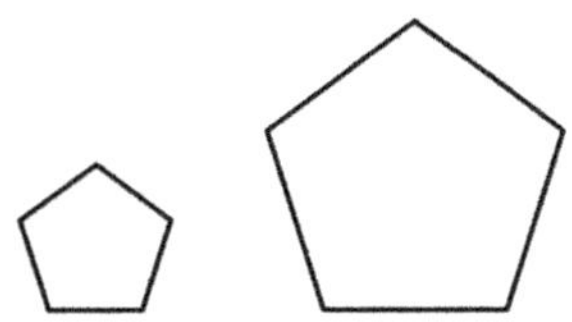

Figure 1 Figure 2

20. The figures 1 and 2 show the regular pentagon of its side length of 1 and 3, respectively. If the area of the figure 1 is 2, then which of the following is the area for the figure 2?

F. 9
G. 16
H. 18
J. 20
K. 24

21. In the xy-plane, the graph of the function f, with equation $f(x) = ax^2 - 3ax - 12$, passes through the point $(-2, 3)$. What is the value of a?

A. $\frac{1}{3}$
B. $\frac{2}{3}$
C. $\frac{3}{2}$
D. $\frac{3}{4}$
E. $\frac{4}{3}$

GO ON TO THE NEXT PAGE.

22. If the ratio of a to b is 2 to 5 and the ratio of b to c is 3 to 7, what is the ratio of a to c ?

F. 15 to 2
G. 10 to 3
H. 5 to 2
J. 5 to 3
K. 6 to 35

23. For all numbers P and Q, where $P \neq Q$, let $P \blacktriangle Q$ be defined as $\frac{P - Q}{P + Q}$. If $4 \blacktriangle t = \frac{2}{5}$, what is the value of t?

A. $\frac{12}{7}$
B. $\frac{4}{5}$
C. 3
D. $\frac{1}{3}$
E. $\frac{5}{2}$

24. If the cube of a certain negative number is equal to a times that number, what is the value of the number in terms of a?

F. $-a$
G. $-a^2$
H. $-\sqrt{a}$
J. $\sqrt{-a}$
K. $-\sqrt{-a}$

25. Which of the following operations could be substituted for Θ, if $a \Theta b = 3a \Theta \frac{b}{3}$ for all positive numbers a and b?

I. $+$
II. $\times$
III. $\div$

A. I only
B. II only
C. III only
D. II and III only
E. I, II, and III

26. If $a + b$ is an irrational number, which of the following statements must be true?

F. a is an irrational number.
G. b is an irrational number.
H. Both a and b are irrational numbers.
J. Either a or b is an irrational number.
K. $a \times b$ is an irrational number.

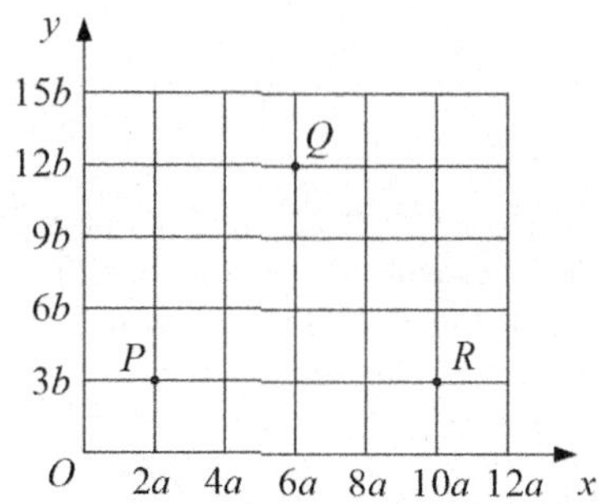

27. On the coordinate system above, what is the area of $\triangle PQR$?

A. $24ab$
B. $30ab$
C. $36ab$
D. $48ab$
E. $52ab$

28. $\frac{317}{100{,}000{,}000}$ can be expressed as 3.17×10^x in scientific notation. What is the value of x?

F. -10
G. -9
H. -8
J. -7
K. -6

29. If n divided by one-half is 35, what is the value of $2n - 10$?

A. 20
B. 25
C. 30
D. 35
E. 40

30. If $x + y = 4$, what is the average (arithmetic mean) of x, y and 7?

F. 3
G. 3.33
H. 3.67
J. 4.33
K. 5

31. The average (arithmetic mean) of seven numbers is equal to the median of the numbers. If the seven numbers are 3, 6, 9, 8, 9, 5 and x, what is one possible value of x?

A. 9
B. 10
C. 12
D. 15
E. 16

GO ON TO THE NEXT PAGE.

32. A printer operating at a constant rate printed 1,000 pages in $83\frac{1}{3}$ minutes. If the printer was continually in operation during this time period, what percent of the 1,000 pages was printed in 1 hour?

F. 66%
G. 67%
H. 69%
J. 70%
K. 72%

33. If $6x - 7y = 5$ and $3x - 13y = (-)7$, what is the value of $x + 2y$?

A. 4
B. 6
C. 7
D. 9
E. 12

34. Suppose n is an integer multiple of 3 that is greater than 50 and less than 90. If n is divided by 9, the remainder must be either 3 or 6. What is one possible value of n that proves that the statement above is NOT true?

F. 53
G. 63
H. 64
J. 71
K. 82

20, 30, 60, 150, 420, x

35. In the sequence above, every term following the first term is formed by multiplying the previous term by a and then subtracting c, where a and c are positive integers. What is the value of x?

A. 1215
B. 1225
C. 1230
D. 1240
E. 1255

36. Paul walked a distance of 2 miles in 35 minutes. If it took 15 minutes for him to walk the first mile, what was his speed, as in miles per hour, for the second mile?

F. 2.7 mile/hr.
G. 2.9 mile/hr.
H. 3.0 mile/hr.
J. 3.2 mile/hr.
K. 3.3 mile/hr.

PATRICK FAMILY'S EXPENSES FOR 2010

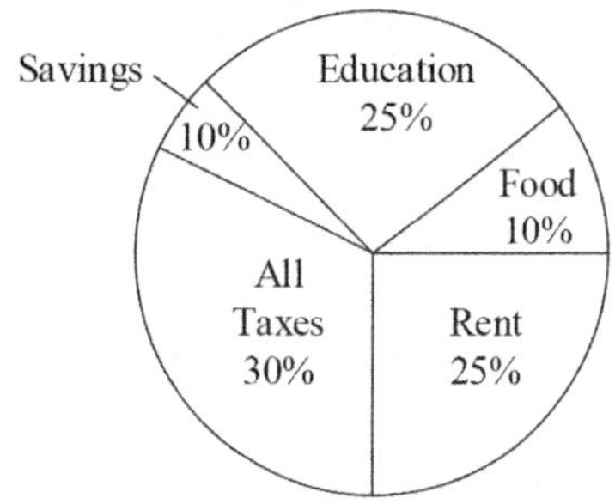

37. The chart above shows the Patrick family's expenses for 2010. The Patrick had an income of \$60,000 in 2010. In 2011, there was 15% increase in Patrick' income. The family spent the same dollar amounts in the three categories of "Rent," "Food," and "Education" in 2011 as they did in 2010. If their taxes increased to 35 percent of their 2011 income, how many dollars went into savings in 2011? (Disregard the \$ sign when gridding your answer)

A. \$2000
B. \$2500
C. \$2800
D. \$8850
E. \$6500

38. Let $a * b$ be defined by $a * b = \frac{5a - 3b}{b}$ for all positive real numbers a and b. What is the value of $2 * (3 * 4)$?

F. 9
G. $\frac{31}{3}$
H. 12
J. $\frac{41}{3}$
K. 24

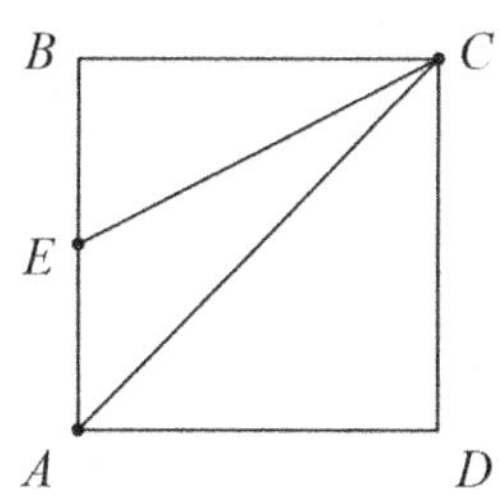

39. In square $ABCD$ above, E is the midpoint of side AB, and let $AE = 1$. What is the perimeter of ΔAEC?

A. $1 + \sqrt{2} + 2\sqrt{3}$
B. $1 + \sqrt{2} + 2\sqrt{5}$
C. $1 + \sqrt{3} + 2\sqrt{2}$
D. $1 + \sqrt{3} + 2\sqrt{5}$
E. $1 + \sqrt{5} + 2\sqrt{2}$

GO ON TO THE NEXT PAGE.

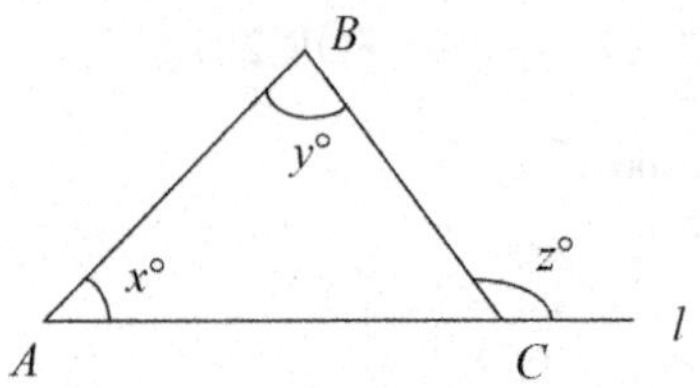

Note: Figure not drawn to scale.

40. Given $\triangle ABC$ and line l shown above, which of the following must be true?

F. The product xyz is 5400.
G. The smallest of x, y and z is y.
H. The largest of x, y and z is y.
J. The sum of x and y is z.
K. If x is 60, then y is 30 and z is 120.

41. The average (arithmetic mean) of seven numbers is 15. If the average of three of them is 11, then what is the average of the other four?

A. 10
B. 10.4
C. 16
D. 18
E. 20

42. Let the operation Δ be defined by $a \Delta b = \frac{ab}{a-b}$ for all numbers a and b, where $a \neq b$. If $2\Delta 3 = 3\Delta x$, what is the value of x?

F. 4
G. 5
H. 6
J. 7
K. 8

43. If 40 is the median of seven numbers 53, 36, 40, 71, 39, 65 and x, the value of x could be any of the following EXCEPT

A. 29
B. 35
C. 39
D. 40
E. 42

44. For all values of a, let $\boxed{a}$ be defined as $\boxed{a} = 3a - 4$. Which of the following equals $\boxed{4} - \boxed{3}$?

F. 3
G. 4
H. 5
J. 6
K. 7

45. If $x^{10} = k^3$ and $x^7 = \frac{8}{27}$, which of the following is an expression for x in terms of k?

A. $\frac{3}{2}k$
B. $\frac{2}{3}k$
C. $2k$
D. $3k$
E. $6k$

46. Judy won a prize of 600 dollars. If she keeps all of her winnings, she has to pay $\frac{1}{3}$ of her winnings in taxes. If she donates 100 dollars of her winnings to charity, then she will pay $\frac{1}{4}$ of the remaining amount in taxes. How much dollars she saves in tax amount, as in without a donation?

F. 100
G. 150
H. 300
J. 120
K. 75

47. If $|x - 2|$ is less than y, which of the following must be true?
I. $x + y > 2$
II. $x - y > 2$
III. $x + y > 2$ and $x - y < 2$

A. I only
B. II only
C. III only
D. I and III
E. II and III

48. Four women are to be seated in a row, and two of them must be next to each other. How many different arrangement are possible?

F. 6
G. 8
H. 12
J. 18
K. 24

49. If $(x + y)^2 = 35$ and $(x - y)^2 = 15$, what is the value of xy?

A. 10
B. 5
C. 7
D. 3
E. 15

GO ON TO THE NEXT PAGE.

50. If $x < y < 0$, which of the following is greatest?

F. $x + y$
G. $x - y$
H. $\frac{y}{x}$
J. 0
K. $\frac{x}{y}$

51. If n is a positive integer, which of the following CANNOT be the units digit of 7^n?

A. 1
B. 3
C. 5
D. 7
E. 9

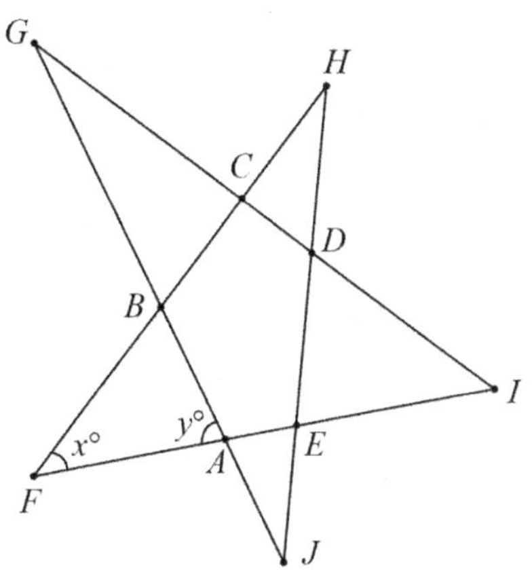

(Note: Figure not drawn to scale.)

52. Given the figure above, suppose that the ⌂ $ABCDE$ is a regular pentagon. Which one of the following is the sum of the angles x and y?

F. 110°
G. 90°
H. 120°
J. 108°
K. 160°

53. Which of the following CANNOT be written as the sum of two prime numbers?

A. 7
B. 8
C. 9
D. 10
E. 11

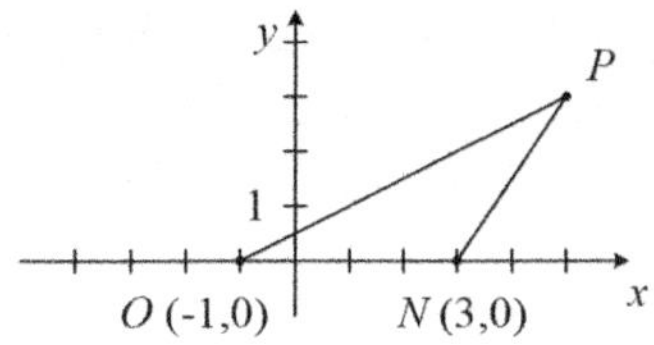

54. In the figure above, ΔOPN has its area equal to 12. Then, which of the following can be the coordinates of the point P?

F. (5, 8)
G. (5, 6)
H. (4, 7)
J. (5, 5)
K. (6, 8)

55. What is the range of $y = 3\cos(\theta + \frac{\pi}{2})$?

A. $-3 < y < 3$
B. $0 \le y$
C. $-\frac{1}{3} \le y \le \frac{1}{3}$
D. $-1 \le y \le 1$
E. $-3 \le y \le 3$

56. If $f(x) = x^2$ and $g(x) = \sqrt{2x+1}$, what is $g(f(2)) + f(1)$?

F. 3
G. 4
H. 6
J. 10
K. $(\sqrt{2} + 2)^2 + 1$

57. If $-2 \le x \le 2$ and $f(x) = |x^2 - 4|$, then what is the range of f?

A. $y \ge 0$
B. $0 \le y \le 7$
C. $5 \le y \le 7$
D. $0 \le y \le 4$
E. $y \le 9$

58. What is the domain of the function: $f(x) = \sqrt{x^2 - 3x + 2}$?

F. $x < 1, x > 2$
G. $-4 < x < 8$
H. $x \ne -4, 0, 3$
J. $x > 0$
K. $x \ne 0, 8$

GO ON TO THE NEXT PAGE.

59. In an arithmetic sequence a_n, $a_2 = 4$ and $a_7 = 14$. What is a_{35}?

A. 45
B. 60
C. 70
D. 82
E. 96

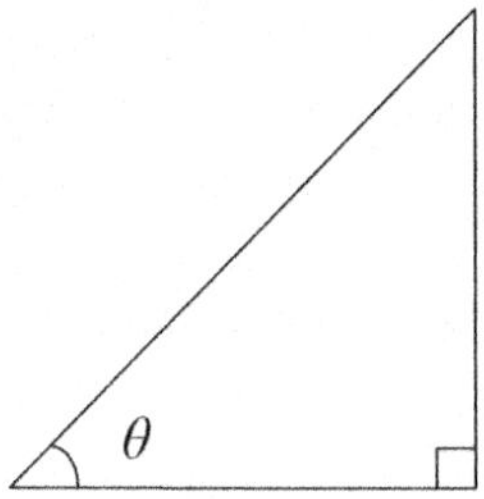

Figure 1

60. In Figure 1, if $\sin\theta = 0.5$ then what is the value of $\tan\theta$?

F. - 1
G. $\frac{1}{3}$
H. $\frac{1}{2}$
J. $\frac{1}{\sqrt{3}}$
K. 1

END OF TEST 3
STOP! DO NOT TURN THE PAGE UNTIL TOLD TO DO SO.
DO NOT RETURN TO THE PREVIOUS TEST.

Answer Key

Model Test No. 03

1	E	16	H	31	E	46	K
2	G	17	D	32	K	47	C
3	D	18	G	33	A	48	H
4	J	19	D	34	G	49	B
5	B	20	H	35	C	50	K
6	J	21	C	36	H	51	C
7	B	22	K	37	D	52	J
8	K	23	A	38	G	53	E
9	D	24	H	39	E	54	G
10	G	25	B	40	J	55	A
11	B	26	J	41	D	56	G
12	H	27	C	42	H	57	D
13	B	28	K	43	E	58	F
14	J	29	B	44	F	59	C
15	C	30	H	45	A	60	J

How to Score the ACT Mathematics Test

When you take an actual ACT Mathematics Test, your answer sheet will be "read" by a scanning machine that will record your responses to each question. Then a computer will compare your answers with the correct answers and produce your raw score. You get one point for each correct answer. For each wrong answer, there is no penalty. Therefore, you must guess and answer all the questions to get a higher score. This raw score is converted to a scaled score that is reported to you and to the colleges you specify.

Finding Your Scale Test Score

STEP 1: Count the number of right answers.
Enter the total here: ____________________

STEP 2: The number you obtained in Step 1 is your raw score.
Look up the conversion table for ACT-Math, and find your score.

Scaled Score Conversion Table
ACT Mathematics Test

Scale Score	Mathematics
36	60
35	58-59
34	56-57
33	55
32	54
31	52-53
30	50-51
29	48-49
28	46-47
27	43-45
26	41-42
25	39-40
24	37-38
23	35-36
22	33-34
21	31-32
20	30
19	27-29
18	25-26
17	22-24
16	18-21
15	15-17
14	12-14
13	09-11
12	08
11	06-07
10	05
09	04
08	-
07	03
06	02
05	-
04	-
03	01
02	-
01	00

1. If the average (arithmetic mean) of 2, 5, and x is 7, what is the average of 5, 8, and x?

A. 5 **B.** 6 **C.** 7 **D.** 8 **E.** 9

$\frac{2+5+x}{3} = 7$

$\therefore 7 + x = 21 \ \therefore x = 14$

Therefore, the average of 5, 8 and x becomes

$\frac{5+8+14}{3} = \frac{27}{3} = 9$

Ans. E

2. If $x + y = 2z$ and $z = 4$, what is the value of $x + y + z$?

F. 8 **G.** 12 **H.** 16 **J.** 24 **K.** 64

$x + y = 2z = 2(4) = 8$

$\therefore x + y + z = (x + y) + z = (8) + 4 = 12$

Ans. G

STUDENTS IN THREE MATH SECTIONS

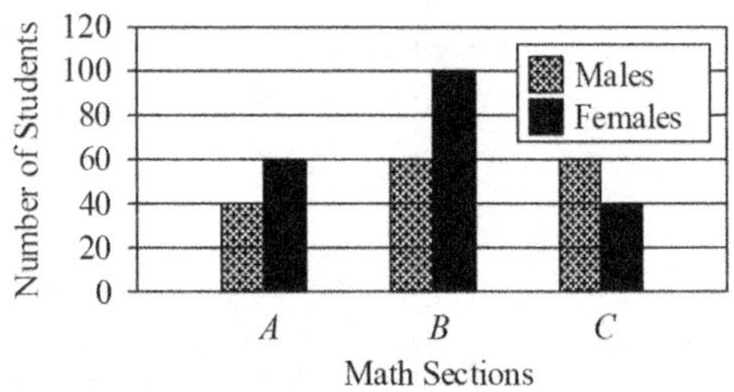

3. The graph above shows the numbers of students in 3 different Math Sections A, B and C. If these sections are to be combined into one section, what will be the ratio of the number of males to the number of females in the new section?

A. 2:3 **B.** 3:2 **C.** 3:5 **D.** 4:5 **E.** 5:4

Let M = number of male combined and F = number of female combined.

Then, M = 40 + 60 + 60 = 160 and F = 60 + 100 + 40 = 200.

$\therefore \frac{M}{F} = \frac{160}{200} = \frac{8}{10} = \frac{4}{5}$

Ans. D

4. Machine A produces a certain auto part at a constant rate of 5 per hour, and machine B produces at a constant rate of 8 per hour. How many more minutes does it take machine A to produce 20 auto parts than it takes machine B?

F. 20 **G.** 60 **H.** 80 **J.** 90 **K.** 120

Let S_A = Speed of machine A and S_B = Speed of machine B,

Then S_A = 5/hr., S_B = 8/hr.

Since time $T = \frac{\text{Work}}{\text{Speed}}$, we get $T_A = \frac{20}{5} = 4$ hrs. or 240 min,

$T_B = \frac{20}{8} = 2\frac{1}{2}$ hrs. or 150 min.

$\therefore T_A - T_B = 240 - 150 = 90$ min.

Ans. J

$\frac{1}{(1)(2)}, \frac{1}{(2)(3)}, \frac{1}{(3)(4)}$

5. The n th term of a sequence is $\frac{1}{(n)(n+1)}$. What is the sum of the first 30 terms of this sequence?

A. 1 **B.** $\frac{30}{31}$ **C.** $\frac{29}{30}$ **D.** $\frac{29}{31}$ **E.** $\frac{14}{30}$

For this problem, notice that $\frac{1}{n(n+1)} = \frac{1}{n} - \frac{1}{n+1}$.

Therefore, $\frac{1}{(1)(2)} + \frac{1}{(2)(3)} + \frac{1}{(3)(4)} + \dots + \frac{1}{(30)(31)}$

$= 1 - \frac{1}{2} + \frac{1}{2} - \frac{1}{3} + \frac{1}{3} - \frac{1}{4} + \frac{1}{4} \cdots \frac{1}{30} + \frac{1}{30} - \frac{1}{31} = 1 - \frac{1}{31} = \frac{30}{31}$

Ans. B

6. Which of the following is an equation of the line in the xy-plane that passes through the point (3, -3) and is perpendicular to the line $y = \frac{1}{2}x - 4$?

F. $y = \frac{1}{2}x + 3$ **G.** $y = -\frac{1}{2}x - 3$ **H.** $y = 2x - 3$

J. $y = -2x + 3$ **K.** $y = -2x - 3$

Note that $m_1m_2 = (-)1$ to be perpendicular, where m_1 and m_2 are the slopes of any two lines in the xy-plane.

Thus, for a given line $y = \frac{1}{2}x - 4$,

an equation of the line that passes through (3, -3) should be,

$y - (-3) = m_2(x - 3)$, where we get, $m_1m_2 = \frac{1}{2}(m_2) = -1$.

$\therefore m_2 = -2 \quad \therefore y + 3 = -2(x - 3)$ or $y = -2x + 3$

Ans. J

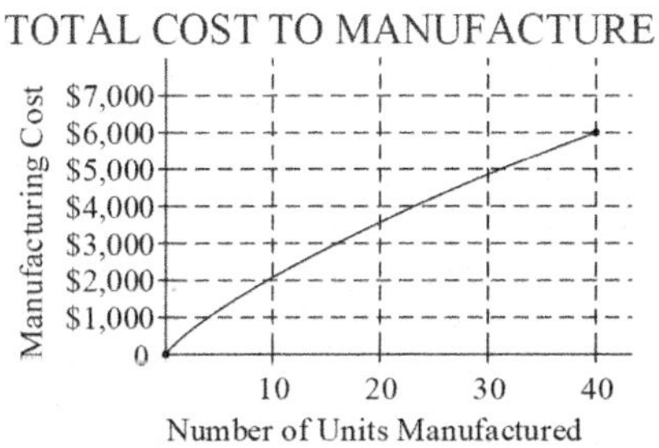

7. The graph above shows the total cost to manufacture a certain type of LCD TV based on the number of units manufactured. What is the approximate cost to manufacture 25 units?

A. $3,500 **B.** $4,400 **C.** $5,200 **D.** $5,700 **E.** $6,100

As seen in the graph, the corresponding cost for 25 units is approximately $4,400.

Ans. B

8. Stained glasses measuring $\frac{1}{2}$ foot by $\frac{1}{3}$ foot are sold in boxes containing 10 each. What is the least number of boxes needed to cover a square window that has dimensions 12 feet by 12 feet?

F. 7 **G.** 16 **H.** 32 **J.** 57 **K.** 87

Each stained glass has area $\frac{1}{2} \times \frac{1}{3} = \frac{1}{6}$ ft^2,

while the total area of the window to cover is $12 \times 12 = 144$ ft^2.

$\therefore 144 \div \frac{1}{6} = 864$ needed. But each box contains 10 stained glasses, therefore, $864 \div 10 = 86.4$ or 87 boxes.

Ans. K

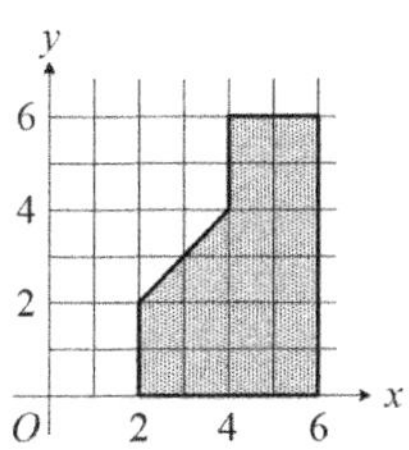

9. What is the area of the shaded part in the figure above?

A. 13 **B.** 14 **C.** 15 **D.** 18 **E.** 22

There are 17 of 1 unit squares plus 2 of $\frac{1}{2}$ unit triangles.

$\therefore 17 + 2\frac{1}{2} = 18$

Ans. D

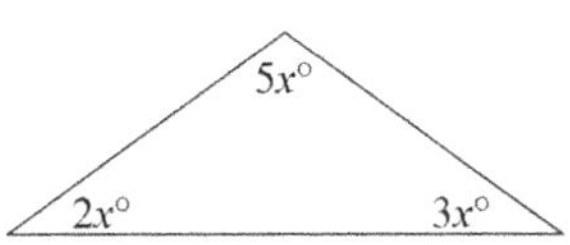

10. In the triangle above, what is the value of x?

F. 20 **G.** 18 **H.** 25 **J.** 15 **K.** 10

$2x + 5x + 3x = 10x = 180°$

$\therefore x = 18°$

Ans. G

11. For expected high temperatures between 80 °F and 100 °F, the number of people in a water park on a given day is modeled by $5(x - 80)(100 - x)$, where x °F is the expected high temperature for that day. The expected high temperatures on July 4th and July 5th are 85 °F and 90 °F, respectively. According to this model, how many more people will be in the water park on July 5th than on July 4th?

A. 5 **B.** 125 **C.** 800 **D.** 900 **E.** 24,700

Let $F(x) = 5(x - 80)(100 - x)$ be the number of people in the water park, for x °F.
Then, on July 4th, $F(85) = 5(85-80)(100-85) = 5(5)(15) = 385$
and on July 5th, $F(90) = 5(90 - 80)(100 - 90) = 5(10)(10) = 500$
$\therefore F(90) - F(85) = 500 - 385 = 125$

Ans. B

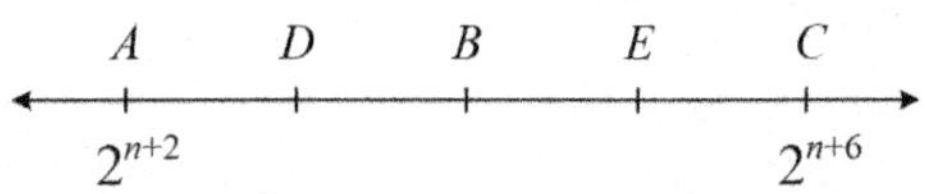

12. On the number line above, point B is the midpoint of $\overline{AC}$ and points D and E are the midpoint of $\overline{AB}$ and $\overline{BC}$ respectively. If n is positive, what is the value at point D?

F. $34 \cdot 2^n$
G. $14 \cdot 2^n$
H. $19 \cdot 2^n$
J. $36 \cdot 2^n$
K. $64 \cdot 2^n$

Much easier way to solve this kind of problem is, Just plug in $n = 1$, and try it.
$A = 2^3$, $C = 2^7$

$\therefore B = \frac{2^3+2^7}{2} = \frac{8+128}{2} = 68$

$\therefore D = \frac{A+B}{2} = \frac{8+68}{2} = 38 = 19 \times 2^1$, which is $19 \cdot 2^n$

Ans. H

13. In a rectangular solid, eight of the edges each have a length of 8 inches and the other four each have a length of 9 inches. What is the volume, in cubic inches, of the rectangular solid?

A. 228 **B.** 576 **C.** 864 **D.** 1,728 **E.** 3,456

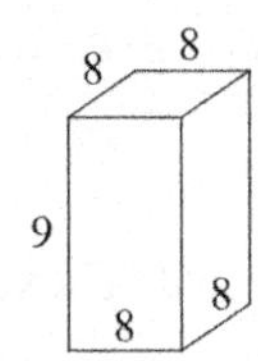

$V = 8 \times 8 \times 9 = 576$

Ans. B

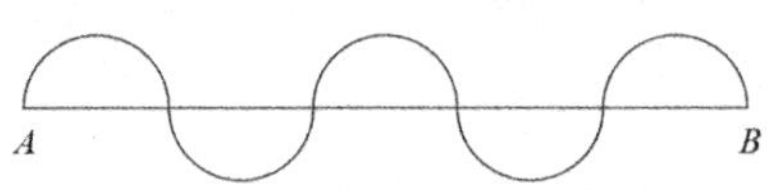

14. In the figure above, the diameters of the 5 semicircles are equal and lie on line segment $\overline{AB}$. If the length of line segment $\overline{AB}$ is 50, what is the area of the 5 semicircles from A to B?

F. 34π **G.** 48π **H.** 40π **J.** 62.5π **K.** 36.5π

The diameter of each semi-circle is, $d = \frac{50}{5} = 10$.
$\therefore r = 5$
$\therefore$ area $= [\pi(5)2 \times \frac{1}{2}] \times 5 = 62.5\pi$

Ans. J

15. If 5 UK quarts are approximately equal to 6 US quarts, which of the following is closest to 1 US quart?

A. 0.5 UK quart
B. 0.7 UK quart
C. 0.8 UK quart
D. 1.2 UK quart
E. 1.8 UK quart

$\frac{5 \text{ UK}}{6 \text{ US}} = \frac{x \text{ UK}}{1 \text{ US}}$

$\therefore x = \frac{5}{6} \approx 0.8$ UK

Ans. C

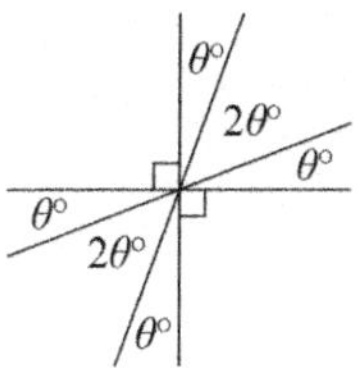

16. In the figure above, what is the value of θ?

F. $\frac{\pi}{4}$ **G.** $\frac{\pi}{6}$ **H.** $\frac{\pi}{8}$ **J.** $\frac{\pi}{10}$ **K.** $\frac{\pi}{12}$

$\theta + 2\theta + \theta = 90° = \frac{\pi}{2}$ $\therefore 4\theta = \frac{\pi}{2}$ $\therefore \theta = \frac{\pi}{8}$

Ans. H

17. If each class period is 45 minutes long, how many class periods are needed to show a 3-hour long video?

A. 1 **B.** 2 **C.** 3 **D.** 4 **E.** 5

$\frac{3 \text{ hrs.} \times 60 \text{ min}}{45 \text{ min}} = 4$

Ans. D

18. The volume of a right circular cone is 18π cubic inches. If the height is twice of the radius of the cone, what is the height of the cone?

F. 4 inches **G.** 6 inches
H. 8 inches **J.** 16 inches
K. 32 inches

$V = \frac{1}{3}\pi r^2 h = 18\pi$
But here, $h = 2r$
$\therefore V = \frac{1}{3}\pi r^2(2r) = \frac{2}{3}\pi r^3 = 18\pi$ $\therefore \frac{2}{3}r^3 = 18$, or $r^3 = 27$
$\therefore r = 3$ $\therefore h = 2r = 2(3) = 6$

Ans. G

19. If $a > 0$, $x^2 + y^2 = a + 5$ and $xy = a - 5$, what is $(x - y)^2$ in terms of a?

A. $-a - 5$ **B.** $a - 5$ **C.** $a + 5$ **D.** $-a + 15$ **E.** $a - 15$

$(x - y)^2 = x^2 - 2xy + y^2 = (x^2 + y^2) - 2xy$
$= (a + 5) - 2(a - 5) = -a + 15$

Ans. D

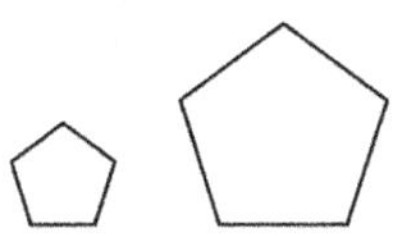

Figure 1 Figure 2

20. The figures 1 and 2 show the regular pentagon of its side length of 1 and 3, respectively. If the area of the figure 1 is 2, then which of the following is the area for the figure 2?

F. 9 **G.** 16 **H.** 18 **J.** 20 **K.** 24

Please note that

Length:	1	→	2	→	3
Area:	1	→	$2^2 = 4$	→	3^2
Volume:	1	→	$2^3 = 8$	→	3^3

$\therefore$ The area of the figure 2 becomes, $2 \times 3^2 = 18$.

Ans. H

21. In the xy-plane, the graph of the function f, with equation $f(x) = ax^2 - 3ax - 12$, passes through the point (-2, 3). What is the value of a?

A. $\frac{1}{3}$ **B.** $\frac{2}{3}$ **C.** $\frac{3}{2}$ **D.** $\frac{3}{4}$ **E.** $\frac{4}{3}$

Since $f(x)$ passes through the points $x = -2$ and $y = 3$, we get $f(-2) = a(-2)^2 - 3a(-2) - 12 = 3$
$\therefore 4a + 6a - 12 = 3$, or $10a = 15$
$\therefore a = \frac{15}{10} = \frac{3}{2}$

Ans. C

22. If the ratio of a to b is 2 to 5 and the ratio of b to c is 3 to 7, what is the ratio of a to c?

F. 15 to 2 **G.** 10 to 3 **H.** 5 to 2 **J.** 5 to 3 **K.** 6 to 35

$\frac{a}{b} = \frac{2}{5}$ and $\frac{b}{c} = \frac{3}{7}$
$\therefore a = \frac{2}{5}b$ and $c = \frac{7}{3}b$ $\therefore \frac{a}{c} = \frac{2b}{5} \div \frac{7b}{3} = \frac{2b}{5} \times \frac{3}{7b} = \frac{6}{35}$

Ans. K

23. For all numbers P and Q, where $P \neq Q$, let $P\blacktriangle Q$ be defined as $\frac{P-Q}{P+Q}$. If $4\blacktriangle t = \frac{2}{5}$, what is the value of t?

A. $\frac{12}{7}$ **B.** $\frac{4}{5}$ **C.** 3 **D.** $\frac{1}{3}$ **E.** $\frac{5}{2}$

Since $P\blacktriangle Q = \frac{(P-Q)}{(P+Q)} = \frac{\text{Diff}}{\text{Sum}}$, we get $4\blacktriangle t = \frac{(4-t)}{(4+t)} = \frac{2}{5}$

Now, cross multiplication gives us $5(4-t) = 2(4+t)$

$\therefore 20 - 5t = 8 + 2t \quad \therefore 12 = 7t$

$\therefore t = \frac{12}{7}$

Ans. A

24. If the cube of a certain negative number is equal to a times that number, what is the value of the number in terms of a?

F. $-a$ **G.** $-a^2$ **H.** $-\sqrt{a}$ **J.** $\sqrt{-a}$ **K.** $-\sqrt{-a}$

Let x be a certain negative number.

Then, $x^3 = a \cdot x \quad \therefore x^2 = a$

$\therefore x = \pm\sqrt{a}$, but since $x < 0$, we have $x = -\sqrt{a}$

Ans. H

25. Which of the following operations could be substituted for $\odot$, if $a \odot b = 3a \odot \frac{b}{3}$ for all positive numbers a and b?

I. +
II. ×
III. ÷

A. I only **B.** II only **C.** III only
D. II and III only **E.** I, II, and III

Let's try $\odot$ be + in (I), then $a + b = 3a + \frac{b}{3}$, not true!

For (II) $\odot = \times$, we get $a \times b = 3a \times \frac{b}{3} = a \times b$, true!

Also, for (III), $a \div b = 3a \div \frac{b}{3} = 9(a \div b)$, not true!

$\therefore$ Ans. B II only.

Ans. B

26. If $a + b$ is an irrational number, which of the following statements must be true?

F. a is an irrational number.
G. b is an irrational number.
H. Both a and b are irrational numbers.
J. Either a or b is an irrational number.
K. $a \times b$ is an irrational number.

For $(a + b)$ to be an irrational number, we only need either one of them is irrational number, not necessarily both.

Ans. J

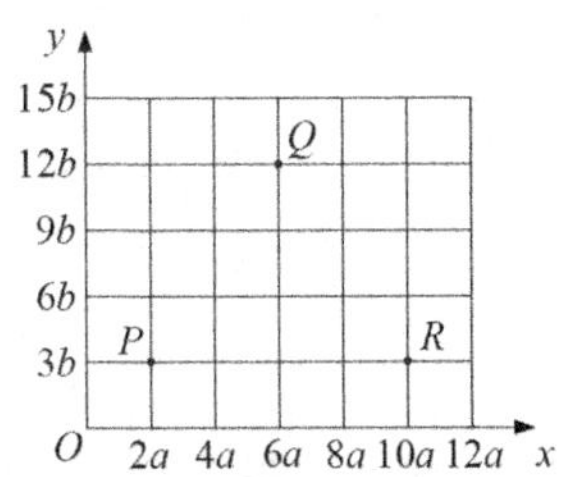

27. On the coordinate system above, what is the area of ΔPQR?

A. $24ab$ **B.** $30ab$ **C.** $36ab$ **D.** $48ab$ **E.** $52ab$

Notice that, for y-axis, it is every $3b$ units, while for x-axis, it is every $2a$ units!!

$\therefore$ area $= \frac{1}{2}$(base)(height) $= \frac{1}{2}(8a)(9b) = 36ab$

Ans. C

28. $\frac{317}{100{,}000{,}000}$ can be expressed as 3.17×10^x in scientific notation. What is the value of x?

F. -10 **G.** -9 **H.** -8 **J.** -7 **K.** -6

$\frac{317}{100{,}000{,}000} = \frac{3.17 \times 10^2}{10^8} = 3.17 \times 10^{-6}$

Ans. K

29. If n divided by one-half is 35, what is the value of $2n$ - 10?

A. 20 **B.** 25 **C.** 30 **D.** 35 **E.** 40

$n \div \frac{1}{2} = n \times \frac{2}{1} = 2n = 35 \quad \therefore 2n - 10 = 35 - 10 = 25$

Ans. B

30. If $x + y = 4$, what is the average (arithmetic mean) of x, y and 7?

F. 3 **G.** 3.33 **H.** 3.67 **J.** 4.33 **K.** 5

$\text{Avg.} = \frac{(x+y)+7}{3} = \frac{(4)+7}{3} = \frac{11}{3}$

Ans. H

31. The average (arithmetic mean) of seven numbers is equal to the median of the numbers. If the seven numbers are 3, 6, 9, 8, 9, 5 and x, what is one possible value of x?

A. 9 **B.** 10 **C.** 12 **D.** 15 **E.** 16

Arranging the number by order, we get 3, 5, 6, 8, 9, 9.
Now, if x is less than 3, then the median becomes 6.
This results, $\text{Avg.} = \frac{x+3+5+6+8+9+9}{7} = 6$
$\therefore x + 40 = 42 \quad \therefore x = 2$, which is one of the possible answers.
You may try x greater than 9, then the median becomes 8.
$\therefore \frac{x+40}{7} = 8 \quad \therefore x + 40 = 56$
$\therefore x = 16$, which is another possible answer!

Ans. E

32. A printer operating at a constant rate printed 1,000 pages in $83\frac{1}{3}$ minutes. If the printer was continually in operation during this time period, what percent of the 1,000 pages was printed in 1 hour?

F. 66% **G.** 67% **H.** 69% **J.** 70% **K.** 72%

Notice that $\text{Speed} = \frac{\text{Work}}{\text{Time}}$.
Therefore, $\text{Speed} = \frac{1000}{83\frac{1}{3}} = \frac{1000}{\frac{250}{3}} = \frac{3000}{250} = 12$ pages/min.
$\therefore$ In 1 hour, we get 12×60 min = 720 pages are printed.
$\therefore \frac{720}{1000} \times 100\% = 72\%$

Ans. K

33. If $6x - 7y = 5$ and $3x - 13y = (-)7$, what is the value of $x + 2y$?

A. 4 **B.** 6 **C.** 7 **D.** 9 **E.** 12

Given $6x - 7y = 5$, eq(1) and $3x - 13y = (-)7$, eq(2),
we get eq(1) – 2 × eq(2) gives us $19y = 19$. $\therefore y = 1$
Now, $y = 1$ into eq(1) gives $x = 2$
$\therefore x + 2y = 2 + 2(1) = 4$

Ans. A

34. Suppose n is an integer multiple of 3 that is greater than 50 and less than 90. If n is divided by 9, the remainder must be either 3 or 6. What is one possible value of n that proves that the statement above is NOT true?

F. 53 **G.** 63 **H.** 64 **J.** 71 **K.** 82

When n is a multiple of 3 between 50 and 90, such as 51, 54, 57, …, 81, 84, 87, the remainder that results from $(n \div 9)$ would be 3, 6 or 0. For example, $54 \div 9 = 6$ with remainder = 0.
Therefore, 54, 63, 72, 81 are the possible answers.

Ans. G

20, 30, 60, 150, 420, x

35. In the sequence above, every term following the first term is formed by multiplying the previous term by a and then subtracting c, where a and c are positive integers. What is the value of x?

A. 1215 **B.** 1225 **C.** 1230 **D.** 1240 **E.** 1255

Given a sequence, 20, 30, 60, 150, 420, x,
we get $30 = 20 \cdot a - c$, eq(1) and
$60 = 30 \cdot a - c$, eq(2)
$\therefore$ eq(2) – eq(1) gives $30 = 10a$
$\therefore a = 3$, which also results, $30 = 20(3) - c \quad \therefore c = 30$.
Therefore, $x = 420(3) - 30 = 1260 - 30 = 1230$

Ans. C

36. Paul walked a distance of 2 miles in 35 minutes. If it took 15 minutes for him to walk the first mile, what was his speed, as in miles per hour, for the second mile?

F. 2.7 mile/hr.
G. 2.9 mile/hr.
H. 3.0 mile/hr.
J. 3.2 mile/hr.
K. 3.3 mile/hr.

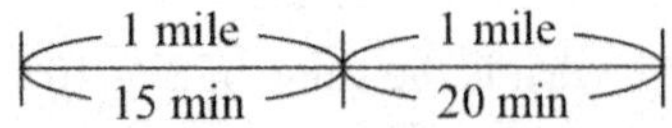

Referring to the above diagram,

we get the speed $= \frac{1 \text{ mile}}{20 \text{ min}} = \frac{1 \text{ mile}}{\frac{1}{3}\text{ hr.}} = 3$ mile/hr.

Ans. H

PATRICK FAMILY'S EXPENSES FOR 2010

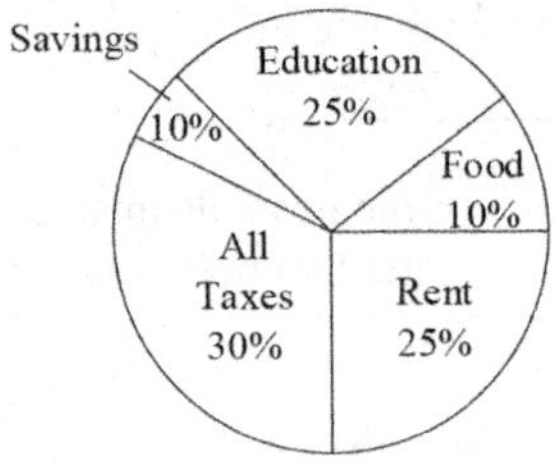

37. The chart above shows the Patrick family's expenses for 2010. The Patrick had an income of \$60,000 in 2010. In 2011, there was 15% increase in Patrick' income. The family spent the same dollar amounts in the three categories of "Rent," "Food," and "Education" in 2011 as they did in 2010. If their taxes increased to 35 percent of their 2011 income, how many dollars went into savings in 2011? (Disregard the \$ sign when gridding your answer.)

A. \$2000 **B.** \$2500 **C.** \$2800 **D.** \$8850 **E.** \$6500

In 2010,
Education = \$60,000 × 0.25 = \$15,000,
Rent = \$60,000 × 0.25 = \$15,000, and
Food = \$60,000 × 0.10 = \$6,000.
They are all fixed, therefore, the same in 2011. Also in 2011, the income is increased to 15%, which is \$60,000 × 1.15 = \$69,000. But the taxes went up to 35% of \$69,000, which is \$69,000 × 0.35 = \$24,150 in 2011. Therefore, the saving in 2011 would be \$69,000 - \$15,000 - \$15,000 - \$6,000 - \$24,150 = \$8,850 in savings.

Ans. D

38. Let $a * b$ be defined by $a * b = \frac{5a - 3b}{b}$ for all positive real numbers a and b. What is the value of $2 * (3 * 4)$?

F. 9 **G.** $\frac{31}{3}$ **H.** 12 **J.** $\frac{41}{3}$ **K.** 24

$3 * 4 = \frac{5(3) - 3(4)}{4} = \frac{3}{4}$

$\therefore 2 * (3 * 4) = 2 * \frac{3}{4} = \frac{5(2) - 3\frac{3}{4}}{\frac{3}{4}} = \frac{10 - \frac{9}{4}}{\frac{3}{4}} = \frac{\frac{31}{4}}{\frac{3}{4}} = \frac{31}{3}$

Ans. G

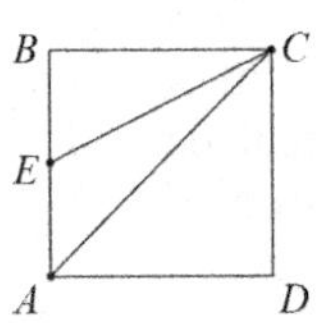

39. In square $ABCD$ above, E is the midpoint of side AB, and let $AE = 1$. What is the perimeter of ΔAEC?

A. $1 + \sqrt{2} + 2\sqrt{3}$
B. $1 + \sqrt{2} + 2\sqrt{5}$
C. $1 + \sqrt{3} + 2\sqrt{2}$
D. $1 + \sqrt{3} + 2\sqrt{5}$
E. $1 + \sqrt{5} + 2\sqrt{2}$

Referring to the above figure, we get the perimeter P of $\Delta AEC = 1 + \sqrt{5} + 2\sqrt{2}$

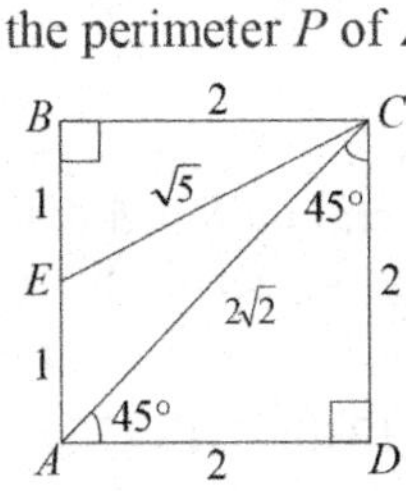

Ans. E

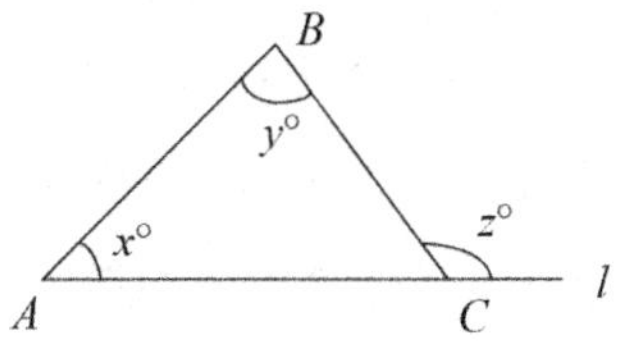

40. Given $\triangle ABC$ and line l shown above, which of the following must be true?

F. The product xyz is 5400.
G. The smallest of x, y and z is y.
H. The largest of x, y and z is y.
J. The sum of x and y is z.
K. If x is 60, then y is 30 and z is 120.

Since angle z is an exterior angle of x and y, we get $x + y = z$.

Ans. J

41. The average (arithmetic mean) of seven numbers is 15. If the average of three of them is 11, then what is the average of the other four?

A. 10 **B.** 10.4 **C.** 16 **D.** 18 **E.** 20

The sum of 7 numbers is $7 \times 15 = 105$.
Also, the sum of 3 numbers is $3 \times 11 = 33$.
$\therefore$ Avg. of the other 4 numbers would be, $\frac{105-33}{4} = 18$.

Ans. D

42. Let the operation $\triangle$ be defined by $a \triangle b = \frac{ab}{a-b}$ for all numbers a and b, where $a \neq b$. If $2\triangle3 = 3\triangle x$, what is the value of x?

F. 4 **G.** 5 **H.** 6 **J.** 7 **K.** 8

$2\triangle3 = 3\triangle x$
$\therefore \frac{2(3)}{2-3} = \frac{3(x)}{3-x}$
$\therefore \frac{6}{-1} = \frac{3x}{3-x}$, or $6(3-x) = -3x$
$\therefore 18 - 6x = -3x$, or $3x = 18$
$\therefore x = 6$

Ans. H

43. If 40 is the median of seven numbers 53, 36, 40, 71, 39, 65 and x, the value of x could be any of the following EXCEPT

A. 29 **B.** 35 **C.** 39 **D.** 40 **E.** 42

36, 39, 40, 53, 65, 71 and x
For the median of seven numbers to be 40, x must be located at 40, or less than 40.
$\therefore$ 42 can't be.

Ans. E

44. For all values of a, let $\boxed{a}$ be defined as $\boxed{a} = 3a - 4$. Which of the following equals $\boxed{4} - \boxed{3}$?

F. 3 **G.** 4 **H.** 5 **J.** 6 **K.** 7

$\boxed{4} = 3(4) - 4 = 8$
$\boxed{3} = 3(3) - 4 = 5$
$\therefore 8 - 5 = 3$

Ans. F

45. If $x^{10} = k^3$ and $x^7 = \frac{8}{27}$, which of the following is an expression for x in terms of k?

A. $\frac{3}{2}k$ **B.** $\frac{2}{3}k$ **C.** $2k$ **D.** $3k$ **E.** $6k$

$x^{10} = x^3 \cdot x^7 = k^3$
But $x^7 = \frac{8}{27}$, thus $x^3 \frac{8}{27} = k^3$,
or $x^3 = \frac{27}{8}k^3 = \frac{27k^3}{8} = \frac{(3k)^3}{2^3} = \left(\frac{3k}{2}\right)^3 = (\frac{3}{2}k)^3$
$\therefore x = \frac{3}{2}k$

Ans. A

46. Judy won a prize of 600 dollars. If she keeps all of her winnings, she has to pay $\frac{1}{3}$ of her winnings in taxes. If she donates 100 dollars of her winnings to charity, then she will pay $\frac{1}{4}$ of the remaining amount in taxes. How much dollars she saves in tax amount, as in without a donation?

F. 100 **G.** 150 **H.** 300 **J.** 120 **K.** 75

The tax without donation is, $600 $\times \frac{1}{3}$ = $200.

Also, the tax with donation is, ($600 - $100) $\times \frac{1}{4}$ = $125.

Therefore, the saving on tax paid will be, $200 - $125 = $75

Ans. K

47. If $|x-2|$ is less than y, which of the following must be true?

I. $x+y>2$
II. $x-y>2$
III. $x+y>2$ and $x-y<2$

A. I only
B. II only
C. III only
D. I and III
E. II and III

$|x-2|<y$
$\therefore -y<x-2<y$, or $2-y<x<2+y$,
which implies, $2<x+y$ and $x-y<2$.

Ans. C

48. Four women are to be seated in a row, and two of them must be next to each other. How many different arrangement are possible?

F. 6
G. 8
H. 12
J. 18
K. 24

Let four women are designated as A, B, C and D.
Then, (A – B) – C – D as we group (A – B) together,
we have (A – B) – (C) – (D), 3 groups.
$\therefore$ 3!, and then there are 2! ways to have (A – B) or (B – A)
$\therefore 3! \times 2! = 12$

Ans. H

49. If $(x+y)^2 = 35$ and $(x-y)^2 = 15$, what is the value of xy?

A. 10
B. 5
C. 7
D. 3
E. 15

$(x+y)^2 = x^2+y^2+2xy = 35$
$\therefore x^2+y^2 = 35-2xy$
Also, $(x-y)^2 = x^2+y^2-2xy = 15$
But here, we replace (x^2+y^2) with $35-2xy$, then we get
$(x^2+y^2)-2xy = (35-2xy)-2xy = 15$, or $35-4xy = 15$.
$\therefore 4xy = 20$
$\therefore xy = 5$

Ans. B

50. If $x<y<0$, which of the following is greatest?

F. $x+y$ **G.** $x-y$ **H.** $\frac{y}{x}$ **J.** 0 **K.** $\frac{x}{y}$

Let $x = -2$ and $y = -1$, then the greatest would be **K**.

Ans. K

51. If n is a positive integer, which of the following CANNOT be the units digit of 7^n?

A. 1 **B.** 3 **C.** 5 **D.** 7 **E.** 9

$7^1 = \boxed{7}$, $7^2 = 4\boxed{9}$, $7^3 = 34\boxed{3}$, $7^4 = 240\boxed{1}$.
Therefore, 7^n has cycle of 7-9-3-1 in every 4 repeating pattern!
$\therefore$ 5 can't be!

Ans. C

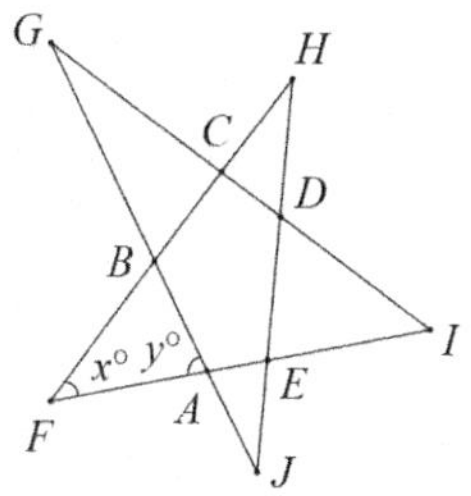

52. Given the figure above, suppose that the ⌂ *ABCDE* is a regular pentagon. Which one of the following is the sum of the angles x and y?

F. 110° **G.** 90° **H.** 120° **J.** 108° **K.** 160°

Since ⌂ *ABCDE* is a regular pentagon,
it has the sum of interior angles, which equals,
$(n-2) \times 180° = (5-2) \times 180° = 540°$
$\therefore \angle FBG = 540 \div 5 = 108°$
Notice that $\angle FBG$ is also an exterior angle of ΔABF,
where $\angle FBG = x + y$.
$\therefore x + y = 108°$

Ans. J

53. Which of the following CANNOT be written as the sum of two prime numbers?

A. 7 **B.** 8 **C.** 9 **D.** 10 **E.** 11

A prime number is the number that has factors of one and itself,
such as 2, 3, 5, 7, …
Now, $7 = 3 + 4$, $8 = 3 + 5$, and $9 = 2 + 7$, $10 = 3 + 7$,
but 11 can't be sum of any two prime numbers.

Ans. E

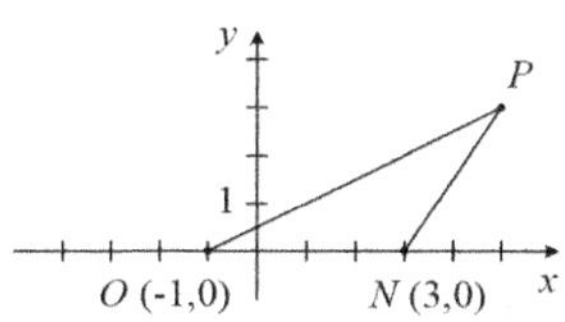

54. In the figure above, ΔOPN has its area equal to 12. Then, which of the following can be the coordinates of the point P?

F. (5, 8) **G.** (5, 6) **H.** (4, 7) **J.** (5, 5) **K.** (6, 8)

ΔOPN has $A = \frac{1}{2}$(base)(height) = 12
Here, the base $= 3 - (-1) = 4$.
$\therefore$ the height should be 6.
The only answer choice that has y-coordinate, $y = 6$, is,

Ans. G

55. What is the range of $y = 3\cos(\theta + \frac{\pi}{2})$?

A. $-3 < y < 3$
B. $0 \leq y$
C. $-\frac{1}{3} \leq y \leq \frac{1}{3}$
D. $-1 \leq y \leq 1$
E. $-3 \leq y \leq 3$

Since $-1 \leq \cos\theta \leq 1$ and also $-1 \leq \cos(\theta + \frac{\pi}{2}) \leq 1$,
we have $-3 \leq 3\cos(\theta + \frac{\pi}{2}) \leq 3$.
Therefore, the range is $-3 \leq y \leq 3$

Ans. A

56. If $f(x) = x^2$ and $g(x) = \sqrt{2x+1}$, what is $g(f(2)) + f(1)$?

F. 3 **G.** 4 **H.** 6 **J.** 10 **K.** $(\sqrt{2} + 2)^2 + 1$

$f(2) = 2^2 = 4$
$\therefore g(f2)) = g(4) = \sqrt{2(4)+1} = \sqrt{9} = 3$
$\therefore g(f(2)) + f(1) = 3 + 1^2 = 4$

Ans. G

57. If $-2 \le x \le 2$ and $f(x) = |x^2 - 4|$, then what is the range of f?

A. $y \ge 0$
B. $0 \le y \le 7$
C. $5 \le y \le 7$
D. $0 \le y \le 4$
E. $y \le 9$

The graph of $f(x) = x^2 - 4$

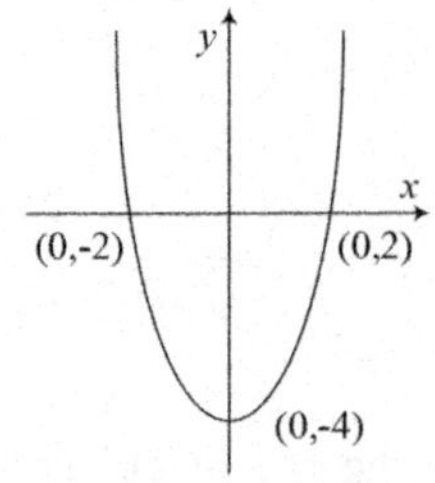

But we have absolute value of $(x^2 - 4)$,
thus we need to flip the curve under x-axis, $f(x) = |x^2 - 4|$.

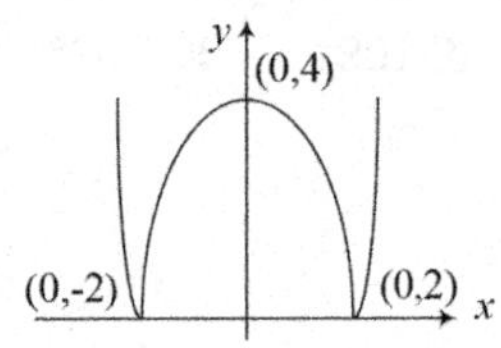

So the range should be, $0 \le y \le 4$

Ans. D

58. What is the domain of the function: $f(x) = \sqrt{x^2 - 3x + 2}$?

F. $x < 1, x > 2$ **G.** $-4 < x < 8$
H. $x \ne -4, 0, 3$ **J.** $x > 0$ **K.** $x \ne 0, 8$

Since inside of radical function must be greater or equal to zero,
$x^2 - 3x + 2 \ge 0$ or $(x-2)(x-1) \ge 0$
That is, $x < 1, x > 2$

Ans. F

59. In an arithmetic sequence a_n, $a_2 = 4$ and $a_7 = 14$. What is a_{35}?

A. 45
B. 60
C. 70
D. 82
E. 96

In arithmetic sequence, the general term a_n is given by
$a_n = a_1 + (n-1)d$
$\therefore a_2 = a_1 + 1d = 4$
$a_7 = a_1 + 6d = 14$
$\therefore a_7 - a_2 = 5d = 14 - 4 = 10 \quad \therefore d = 2$
Therefore, back into a_2, we get $a_2 = 4 = a_1 + 2 \quad \therefore a_1 = 2$
Now, $a_{35} = a_1 + 34d = 2 + 34(2) = 70$

Ans. C

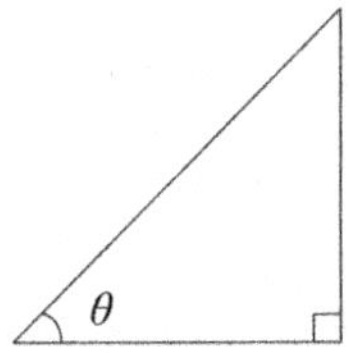

Figure 1

60. In Figure 1, if $\sin\theta = 0.5$ then what is the value of $\tan\theta$?

F. -1 **G.** $\frac{1}{3}$ **H.** $\frac{1}{2}$ **J.** $\frac{1}{\sqrt{3}}$ **K.** 1

If $\sin\theta = \frac{1}{2}$,
then we get

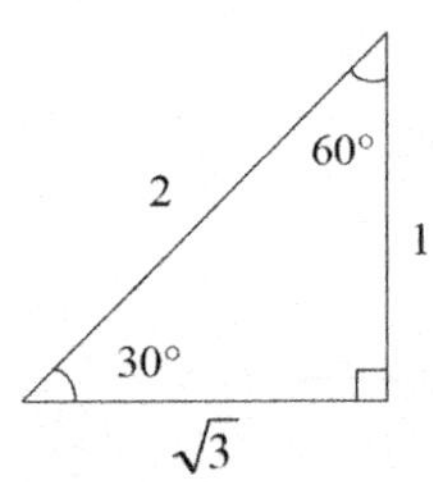

Therefore, $\tan\theta = \frac{1}{\sqrt{3}}$

Ans. J

Model Test No. 04

60 Questions / 60 Minutes

Directions: Solve each problem, choose the correct answer, and then fill in the corresponding oval on your answer document.

Do not linger over problems that take too much time. Solve as many as you can; then return to the others in the time you have left for this test.

You are permitted to use a calculator on this test. You many use your calculator for any problems you choose, but some of the problems may best be done without using a calculator.

Note: Unless otherwise stated, all of the following should be assumed.

1. Illustrative figures are NOT necessarily drawn to scale.
2. Geometric figures lie in a plane.
3. The word line indicates a straight line.
4. The word average indicates arithmetic mean.

Answer Sheet

Model Test No. 04

1	Ⓐ Ⓑ Ⓒ Ⓓ Ⓔ	16	Ⓕ Ⓖ Ⓗ Ⓙ Ⓚ	31	Ⓐ Ⓑ Ⓒ Ⓓ Ⓔ	46	Ⓕ Ⓖ Ⓗ Ⓙ Ⓚ
2	Ⓕ Ⓖ Ⓗ Ⓙ Ⓚ	17	Ⓐ Ⓑ Ⓒ Ⓓ Ⓔ	32	Ⓕ Ⓖ Ⓗ Ⓙ Ⓚ	47	Ⓐ Ⓑ Ⓒ Ⓓ Ⓔ
3	Ⓐ Ⓑ Ⓒ Ⓓ Ⓔ	18	Ⓕ Ⓖ Ⓗ Ⓙ Ⓚ	33	Ⓐ Ⓑ Ⓒ Ⓓ Ⓔ	48	Ⓕ Ⓖ Ⓗ Ⓙ Ⓚ
4	Ⓕ Ⓖ Ⓗ Ⓙ Ⓚ	19	Ⓐ Ⓑ Ⓒ Ⓓ Ⓔ	34	Ⓕ Ⓖ Ⓗ Ⓙ Ⓚ	49	Ⓐ Ⓑ Ⓒ Ⓓ Ⓔ
5	Ⓐ Ⓑ Ⓒ Ⓓ Ⓔ	20	Ⓕ Ⓖ Ⓗ Ⓙ Ⓚ	35	Ⓐ Ⓑ Ⓒ Ⓓ Ⓔ	50	Ⓕ Ⓖ Ⓗ Ⓙ Ⓚ
6	Ⓕ Ⓖ Ⓗ Ⓙ Ⓚ	21	Ⓐ Ⓑ Ⓒ Ⓓ Ⓔ	36	Ⓕ Ⓖ Ⓗ Ⓙ Ⓚ	51	Ⓐ Ⓑ Ⓒ Ⓓ Ⓔ
7	Ⓐ Ⓑ Ⓒ Ⓓ Ⓔ	22	Ⓕ Ⓖ Ⓗ Ⓙ Ⓚ	37	Ⓐ Ⓑ Ⓒ Ⓓ Ⓔ	52	Ⓕ Ⓖ Ⓗ Ⓙ Ⓚ
8	Ⓕ Ⓖ Ⓗ Ⓙ Ⓚ	23	Ⓐ Ⓑ Ⓒ Ⓓ Ⓔ	38	Ⓕ Ⓖ Ⓗ Ⓙ Ⓚ	53	Ⓐ Ⓑ Ⓒ Ⓓ Ⓔ
9	Ⓐ Ⓑ Ⓒ Ⓓ Ⓔ	24	Ⓕ Ⓖ Ⓗ Ⓙ Ⓚ	39	Ⓐ Ⓑ Ⓒ Ⓓ Ⓔ	54	Ⓕ Ⓖ Ⓗ Ⓙ Ⓚ
10	Ⓕ Ⓖ Ⓗ Ⓙ Ⓚ	25	Ⓐ Ⓑ Ⓒ Ⓓ Ⓔ	40	Ⓕ Ⓖ Ⓗ Ⓙ Ⓚ	55	Ⓐ Ⓑ Ⓒ Ⓓ Ⓔ
11	Ⓐ Ⓑ Ⓒ Ⓓ Ⓔ	26	Ⓕ Ⓖ Ⓗ Ⓙ Ⓚ	41	Ⓐ Ⓑ Ⓒ Ⓓ Ⓔ	56	Ⓕ Ⓖ Ⓗ Ⓙ Ⓚ
12	Ⓕ Ⓖ Ⓗ Ⓙ Ⓚ	27	Ⓐ Ⓑ Ⓒ Ⓓ Ⓔ	42	Ⓕ Ⓖ Ⓗ Ⓙ Ⓚ	57	Ⓐ Ⓑ Ⓒ Ⓓ Ⓔ
13	Ⓐ Ⓑ Ⓒ Ⓓ Ⓔ	28	Ⓕ Ⓖ Ⓗ Ⓙ Ⓚ	43	Ⓐ Ⓑ Ⓒ Ⓓ Ⓔ	58	Ⓕ Ⓖ Ⓗ Ⓙ Ⓚ
14	Ⓕ Ⓖ Ⓗ Ⓙ Ⓚ	29	Ⓐ Ⓑ Ⓒ Ⓓ Ⓔ	44	Ⓕ Ⓖ Ⓗ Ⓙ Ⓚ	59	Ⓐ Ⓑ Ⓒ Ⓓ Ⓔ
15	Ⓐ Ⓑ Ⓒ Ⓓ Ⓔ	30	Ⓕ Ⓖ Ⓗ Ⓙ Ⓚ	45	Ⓐ Ⓑ Ⓒ Ⓓ Ⓔ	60	Ⓕ Ⓖ Ⓗ Ⓙ Ⓚ

MATHEMATICS TEST

60 Minutes – 60 Questions

DIRECTIONS: Solve each problem, choose the correct answer, and then fill in the corresponding oval on your answer document.

Do not linger over problems that take too much time. Solve as many as you can; then return to the others in the time you have left for this test.

You are permitted to use a calculator on this test. You many use your calculator for any problems you choose, but some of the problems may best be done without using a calculator.

Note: Unless otherwise stated, all of the following should be assumed.

1. Illustrative figures are NOT necessarily drawn to scale.
2. Geometric figures lie in a plane.
3. The word line indicates a straight line.
4. The word average indicates arithmetic mean.

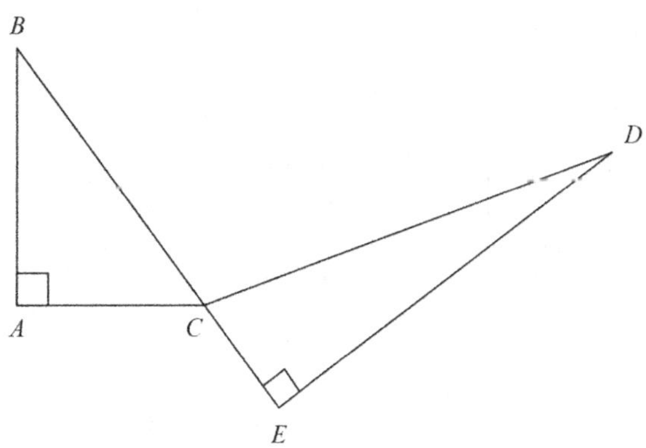

Note: Figure not drawn to scale.

1. In the figure above, point C lies on $\overline{BE}$. If $AB = 8$, $AC = 6$, $CD = 13$, and $DE = 12$, what is the length of $\overline{BE}$?

   A. 11
   B. 12
   C. 13
   D. 14
   E. 15

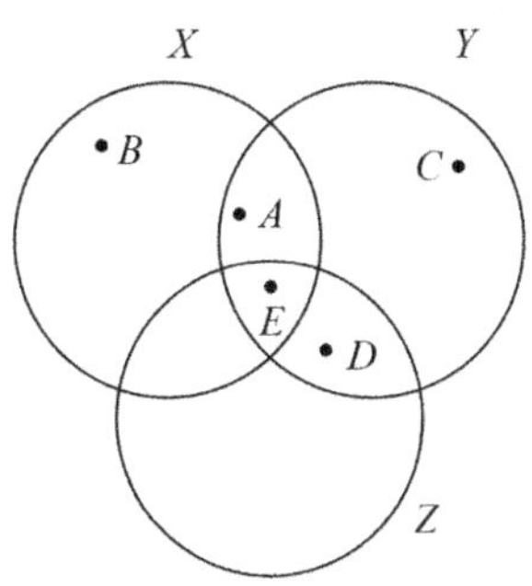

2. In the figure above, which of the labeled points is inside circle X and circle Y, but not in circle Z?

   F. A
   G. B
   H. C
   J. D
   K. E

3. If $\frac{2a^2b}{ab^2} = \frac{3}{2}$, then $\frac{a}{b} =$

   A. 3
   B. $\frac{3}{2}$
   C. $\frac{4}{3}$
   D. $\frac{3}{4}$
   E. $\frac{1}{3}$

4. The measure of the smallest angle in a certain triangle is one fifth the sum of the measures of the remaining angles. What is the measure of the smallest angle?

   F. 20°
   G. 30°
   H. 36°
   J. 45°
   K. 50°

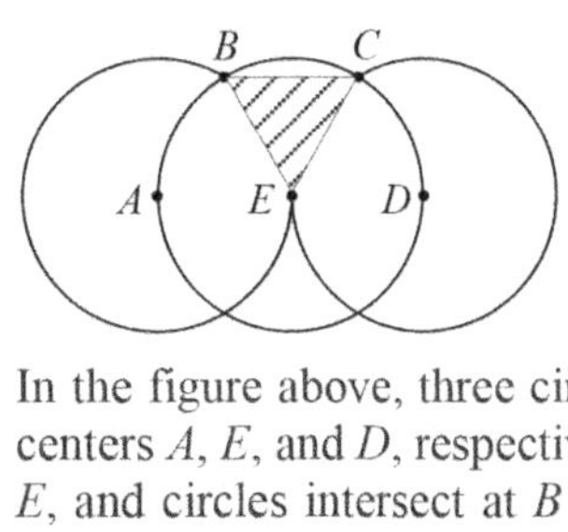

5. In the figure above, three circles, each with radius 6, have centers A, E, and D, respectively. Two circles are tangent at E, and circles intersect at B and C, as shown. What is the area of triangle EBC shown above?

   A. 18
   B. $18\sqrt{3}$
   C. $6\sqrt{3}$
   D. $9\sqrt{3}$
   E. 9

GO ON TO THE NEXT PAGE.

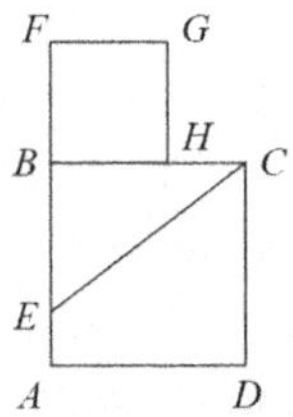

6. In the figure above, $ABCD$ and $BFGH$ are squares, $4BE = 3AB$ and $FE = CE$. If $BF = 1$, what is the length of $\overline{AD}$?

   F. $\frac{2}{3}$
   G. $\frac{3}{2}$
   H. 2
   J. $\frac{5}{2}$
   K. 3

7. If $5 - 3x = 5$, what is the value of $(5 - 3x)(5 + 3x)$?

   A. 5
   B. 10
   C. 15
   D. 20
   E. 25

8. The measures of the three angles of a triangle are $5a°$, $3a°$and $2a°$. The measure of the largest angle of the triangle is how much greater than the measure of the smallest angle?

   F. 15°
   G. 36°
   H. 54°
   J. 72°
   K. 105°

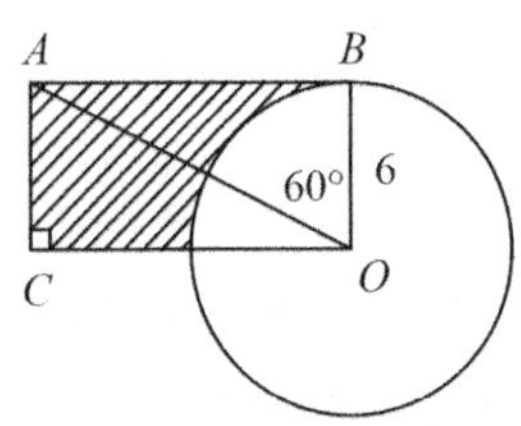

9. In the figure above, the center of the circle is O and $\overline{AB}$ is tangent to the circle at point B. What is the area of the shaded region of the rectangle $ABOC$?

   A. $36\sqrt{3} - 6\pi$
   B. $36\sqrt{3} - 12\pi$
   C. $18\sqrt{3} - 6\pi$
   D. $18\sqrt{3} - 12\pi$
   E. $36\sqrt{3} - 9\pi$

10. In the xy-plane, line l passes through the point (1, 2) and the vertex of the parabola with the equation $y = 2x^2 - 12x + 12$. What is the slope of line l?

   F. -4
   G. $-\frac{1}{4}$
   H. 0
   J. $\frac{1}{4}$
   K. 4

NUMBER OF YEARS OF DRIVING EXPERIENCE

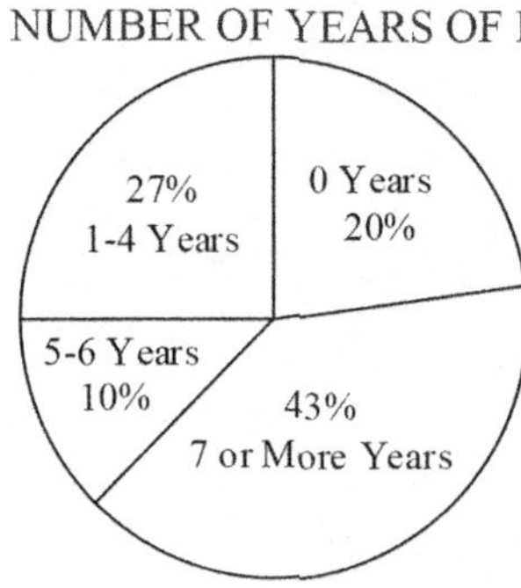

11. The graph above shows the distribution of the number of years of driving experience for a group of 1,000 adults. According to the graph, how many of these adults had 5 or more years of driving experience?

   A. 530
   B. 620
   C. 740
   D. 180
   E. 390

12. If 3.38 is rounded to the nearest tenth and the result is tripled, what is the final result?

   F. 10.4
   G. 10.3
   H. 10.2
   J. 10.1
   K. 10.0

13. At a book sale, all paperback books cost d dollars each and all hardback books cost $3d$ dollars each. If Emily paid $13d$ dollars for 7 books at this sale, how many hardback books did she buy?

   A. None
   B. One
   C. Two
   D. Three
   E. Four

GO ON TO THE NEXT PAGE.

14. For which of the following functions is it true that $f(-x) = f(x)$ for all values of x?

F. $f(x) = x + 4$
G. $f(x) = x^2 + 4$
H. $f(x) = x^3 + 4$
J. $f(x) = x^2 + x$
K. $f(x) = x^3 + x$

15. Which of the following represents half the difference between $4x$ and 14?

A. $7 - 4x$
B. $3x - 7$
C. $2x - 14$
D. $2x - 7$
E. $14 - 2x$

GROWTH OF A POPULATION OF BACTERIA

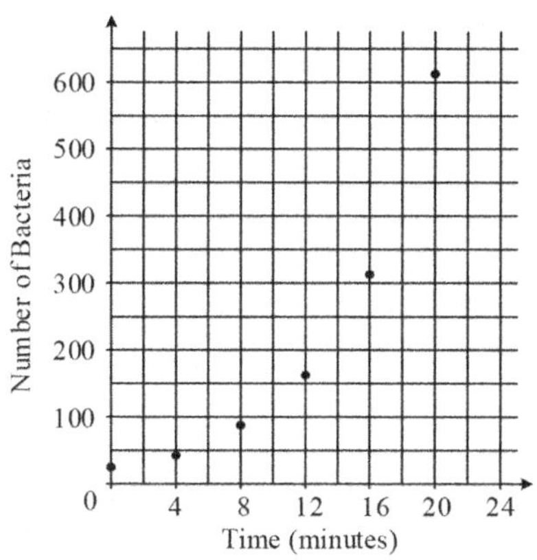

16. The number of bacteria in a dish at various times from the start of an experiment is shown in the graph above. Based on the pattern indicated by the graph, which of the following is closest to the number of minutes from the start of the experiment to the time when there were 450 bacteria in the dish?

F. 6
G. 10
H. 14
J. 18
K. 22

17. If j and k are integers and $0 \leq j \leq k \leq 6$, which of the following is not a possible value of jk?

A. 9
B. 14
C. 16
D. 18
E. 20

18. If the average (arithmetic mean) of x, y and 12 is 8, what is the average of x and y?

F. 2
G. 3
H. 6
J. 7
K. It cannot be determined from the information given

19. If $2 + \sqrt{x} = 4$, then which of the following is $(x + 2)^2$?

A. 36
B. 24
C. 18
D. 12
E. 9

20. If $t = 2x - 3$ and $t = y - 7$, which of the following must be true?

F. $t = x$
G. $x = y$
H. $x < y$
J. $x > y$
K. infinitely many solutions

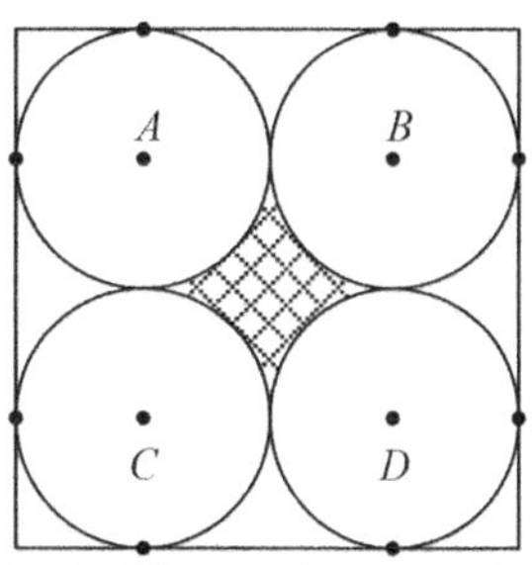

21. In the figure above, each of the points A, B, C and D is the center of a circle of radius 1. The points where the circles are tangent to each other and to the square are marked. What is the area of the shaded part?

A. $8 - \pi$
B. $4 - \pi$
C. π
D. 4π
E. 2π

22. If n is a positive integer, which of the following is equivalent to $9^n + 9^n$?

F. $2 \cdot 3^{2n}$
G. 3^{4n}
H. $(2 \cdot 3 \cdot 3)^n$
J. $(2 \cdot 3)^{2n}$
K. $(3 \cdot 3 \cdot 3 \cdot 3)^{2n}$

GO ON TO THE NEXT PAGE.

23. If a number is chosen at random from the set {-8, -6, -4, -2, 0, 2, 4, 6, 8}, what is the probability that it is a member of the solution set of both $|x-1|<7$ and $|2x+5|>3$?

A. 0

B. $\frac{2}{3}$

C. $\frac{4}{9}$

D. $\frac{5}{9}$

E. $\frac{7}{9}$

24. $(2^x \cdot 2^{3x})^{2x} =$

F. 2^{6x^3}
G. 2^{6x}
H. 2^{8x^2}
J. 2^{8x}
K. 2^{16x^2}

25. The eight letters *a*, *b*, *c*, *d*, *e*, *f*, *g* and *h* represent the eight integers 2 through 9, not necessarily in that order. If *b* is the product of prime numbers *e* and *h*, which of the following could NOT be the value of *f*?

A. 5
B. 6
C. 7
D. 8
E. 9

26. The radius of a wheel is 6 feet. If the wheel rolls 5 miles without slipping, how many revolutions will it make? (1 mile = 5,280 feet)

F. 400
G. 500
H. 600
J. 700
K. 900

27. If $a-3b$ is 5 more than $2a-2b$, the value of which of the following expressions can be determined?

A. a
B. b
C. $a-3b$
D. $a-b$
E. $a+b$

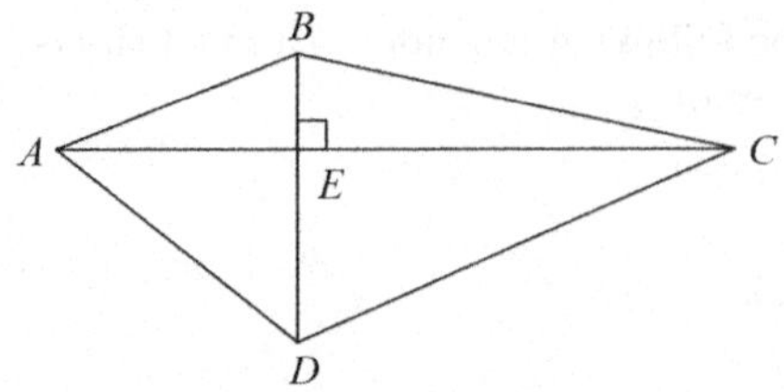

28. In $ABCD$ above, $AC = 15$, $BD = 8$ and $\overline{AC} \perp \overline{BD}$. What is the area of quadrilateral $ABCD$?

F. 15
G. 30
H. 45
J. 60
K. 120

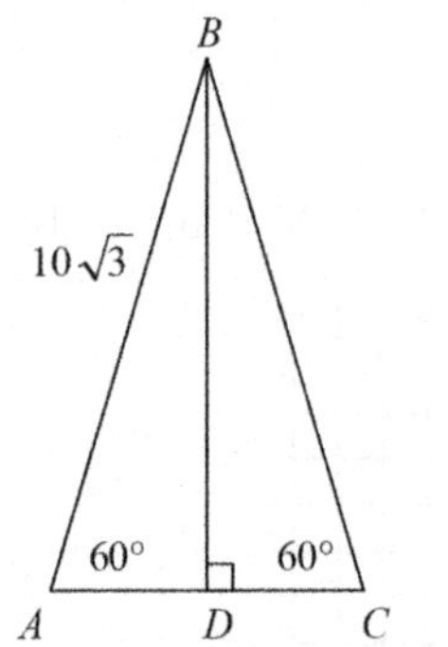

Note: Figure not drawn to scale.

29. In $\triangle ABC$ above, what is the length of BD?

A. $7\sqrt{3}$
B. $8\sqrt{3}$
C. 15
D. 17
E. $11\sqrt{3}$

30. If $2^{2n+1} = 8^5$, what is the value of n?

F. 3
G. 4
H. 5
J. 7
K. 8

31. In a group of 500 students, 350 are studying Spanish, 108 are studying Chinese, and 26 are studying Latin. What is the least possible number of these students that might NOT be studying any of these languages?

A. 10
B. 12
C. 13
D. 14
E. 16

GO ON TO THE NEXT PAGE.

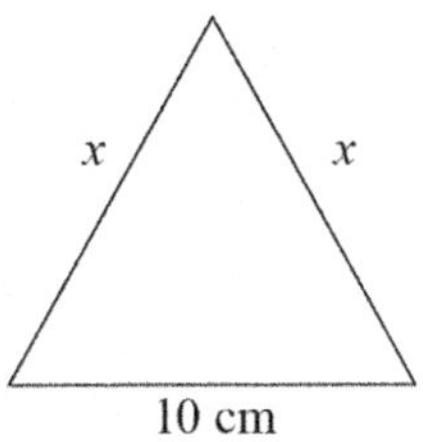

Note: Figure not drawn to scale.

32. If the perimeter of the triangle above is 36 centimeters, what is its area, in square centimeters?

F. 48
G. 54
H. 58
J. 60
K. 64

33. What is the number that satisfies the following three conditions?

- It is 4 digit integer less than 1200.
- The sum of its digits is 12.
- Its tens and units digits are the same.

A. 1074
B. 1065
C. 1155
D. 1164
E. 1173

34. Point C is on line segment AB such that $\frac{AC}{AB}=\frac{3}{5}$. If the length of AC is 9, what is the length of CB?

F. 4.8
G. 5
H. 5.2
J. 6
K. 6.4

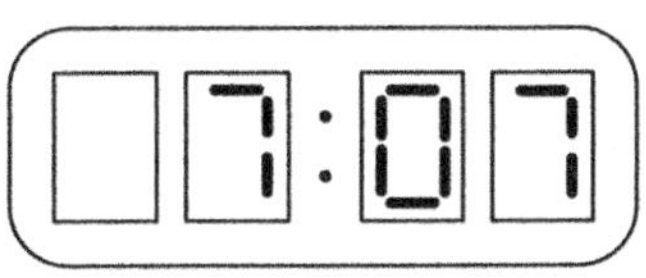

35. The face of a 12-hour digital clock, pictured above, shows one example of a time at which reading the digits from left to right is the same as reading the digits from right to left. How many such times would be shown on this clock in a 12-hour period starting at 12 midnight?

A. 49
B. 52
C. 54
D. 55
E. 57

36. The first term of a sequence of numbers is 12. If each term after the first is 4 more than 3 times the preceding term, what is the fourth term of this sequence?

F. 364
G. 372
H. 376
J. 384
K. 396

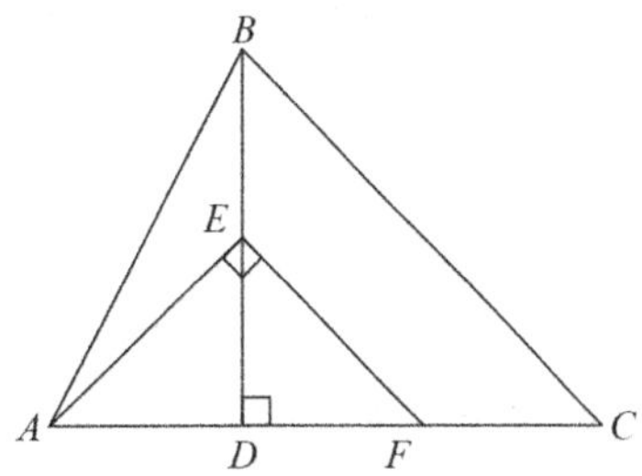

Note: Figure not drawn to scale.

37. In ΔABC above, $AD = ED$, $BC = 2EF$ and $\overline{BC} \parallel \overline{EF}$. What is the value of the fraction $\frac{\text{area } \square AEF}{\text{area } \square ABC}$?

A. $\frac{2}{9}$
B. $\frac{1}{3}$
C. $\frac{4}{9}$
D. $\frac{1}{2}$
E. $\frac{5}{9}$

38. If $-1.7 < \frac{n}{7} < -1.2$, where n is an integer, what is one possible value of $|n|$?

F. 7
G. 8
H. 10
J. 12
K. 15

39. Ali, Ben, Chris, David and Emily are each to be assigned to 1 of 2 teams. How many different assignments of these 5 people are possible if at least 1 person must be assigned to each team?

A. 5
B. 10
C. 15
D. 20
E. 30

GO ON TO THE NEXT PAGE.

40. If p, r, and s are any three consecutive positive integers, which of the following must be even?

I. prs
II. $p + r + s$
III. $p^2 + r^2 + s^2$

F. I only
G. II only
H. III only
J. I and III only
K. I, II, and III

41. If x is a percent of b and if b is c percent of y, then x is what percent of y? (Assume all variables are positive.)

A. $\frac{a+b+c}{100}\%$
B. $\frac{ab}{100}\%$
C. $\frac{bc}{100}\%$
D. $\frac{ac}{100}\%$
E. $\frac{abc}{100}\%$

42. When the number x is multiplied by 7, the result is the same as when 2 is added to x. What is the value of $4x$?

F. $\frac{3}{4}$
G. 1
H. $\frac{4}{3}$
J. 3
K. 4

5.10100100001000000001...

43. The decimal number above consists of only 1's and 0's to the right of the decimal point. The first 1 is followed by one 0, the second 1 is followed by two 0's, the third 1 is followed by four 0's, the fourth 1 is followed by eight 0's and so on. What is the total number of 0's between the 9th and the 11th 1 in this decimal number?

A. 256
B. 512
C. 768
D. 1536
E. 1792

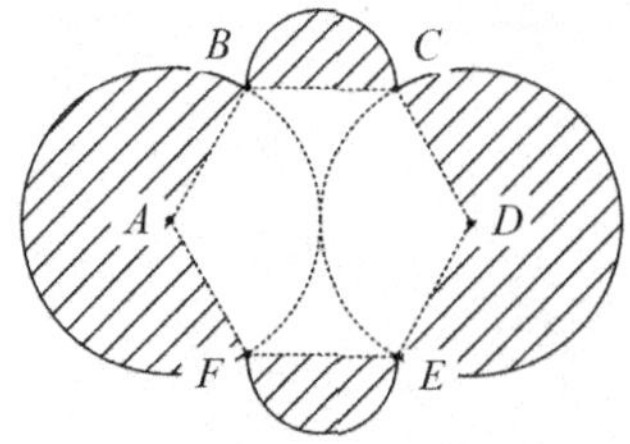

44. In the figure above, $ABCDEF$ is a regular hexagon whose sides are each 6 units long. A and D are the centers of the two large circles, and BC and FE are diameters of the two small semicircles. What is the total area of the figure?

F. 57π
G. 54π
H. 55π
J. 56π
K. 50π

45. There are only white and black marbles in a box. If 4 marbles are white and the probability of choosing a white marble at random from the box is $\frac{1}{4}$, how many black marbles are in the box?

A. 18
B. 12
C. 9
D. 6
E. 2

46. If n and k are positive odd integers, which of the following has the least value?

F. n^k
G. $(-n)^{-k}$
H. $(-n)^{2k}$
J. $(-2n)^{-k}$
K. $(-2n)^{-2k}$

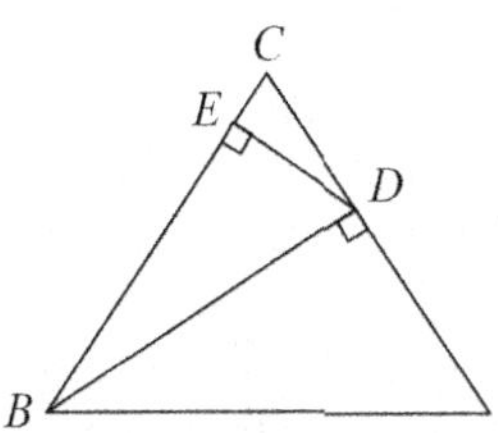

47. In the figure above, ΔABC is an equilateral triangle. What is the ratio of the areas of ΔABC to ΔCDE?

A. 2:1
B. 4:1
C. 6:1
D. 8:1
E. 16:1

GO ON TO THE NEXT PAGE.

48. If a wheel revolves at the rate of $\frac{2}{3}\pi$ per second, how many complete revolutions does it make in one <u>minute</u>?

F. 9,000
G. 150
H. 144
J. 40
K. 20

49. The least integer of a set of consecutive integers is -25. If the sum of these integers is 26, what is the average of these integers?

A. 0
B. $\frac{1}{2}$
C. $\frac{1}{4}$
D. 1
E. $\frac{26}{51}$

50. If a and b are positive integers, which of the following expressions is equivalent to $\frac{(2^a \cdot 2^a)^b}{2^a}$?

F. 2^{a+b}
G. 2^{ab}
H. 2^{2b}
J. $2^{a(ab-1)}$
K. $2^{a(2b-1)}$

51. If a person needs 3 feet of space along the edge of a circular table to be seated comfortably, what is the maximum number of people who can be seated comfortably at a table of radius 7 feet?

A. 12
B. 14
C. 7
D. 21
E. 15

52. The sum of 5 consecutive even integers is 1,500. What is the least of these integers?

F. 326
G. 296
H. 272
J. 300
K. 216

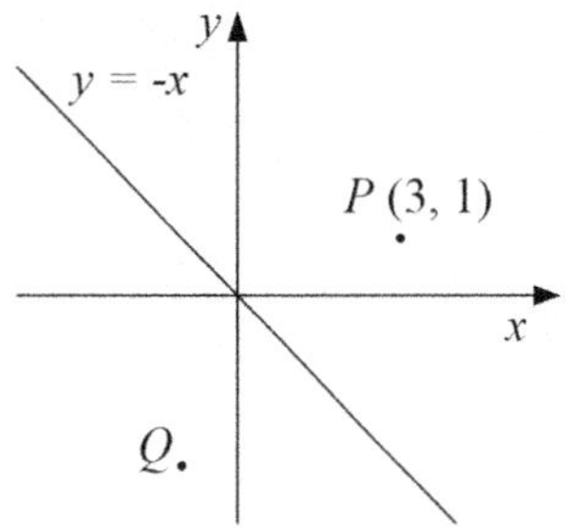

53. In the figure above, point P is the reflection of point Q through the line $y = -x$. What is the coordinates of point Q?

A. (-1, -3)
B. (-1, -2)
C. (-3, -1)
D. $(-\frac{3}{2}, -\frac{1}{2})$
E. (-2, -1)

54. Peter slices a pizza into k equal pieces and eats three pieces. In terms of k, what percent of the pizza is left?

F. $100(k - 3)$ %
G. $\frac{100(k-3)}{k}$ %
H. $\frac{100k}{k-3}$ %
J. $\frac{k-3}{100}$ %
K. $\frac{k-3}{100k}$ %

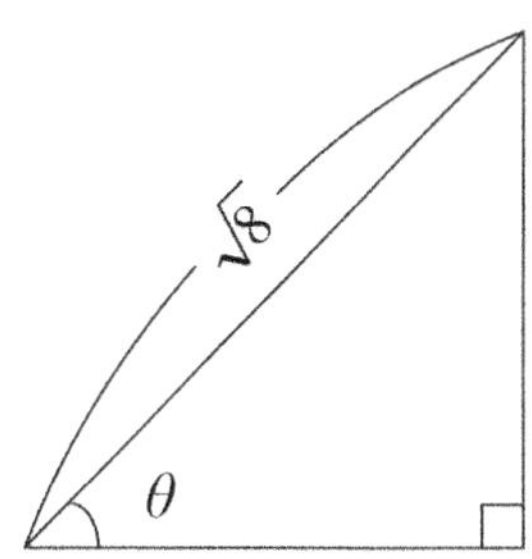

Figure 1

55. In Figure 1, if $\theta = 45°$, what is the area of the triangle?

A. 2
B. 2.36
C. 4
D. 4.72
E. 6

GO ON TO THE NEXT PAGE.

56. If $\sqrt{(x+2)} = 3$, then $x =$

F. -3.5
G. 2
H. 4.2
J. 7
K. 12

57. If $f(7x + 4) = 28x + 16$ for all real x, what is $f(x)$?

A. $x + 5$
B. $7x + 5$
C. $4x$
D. $2x + 5$
E. $2x - 5$

58. If the line $y = mx + b$ intersects the parabola $y = 2x^2 + 3$ at $x = 0$, what is the y-intercept of the line?

F. 8
G. 3
H. 5
J. 4
K. 2.5

59. $f(x) = \left\{ \begin{array}{l} x^2 - 4 \text{ if } x > 2 \\ -2x + 4 \text{ if } x \leq 2 \end{array} \right\}$. What is the range of f?

A. $y \leq 1$
B. $y > 1$
C. $y \geq 0$
D. $y \geq 3$
E. All real numbers

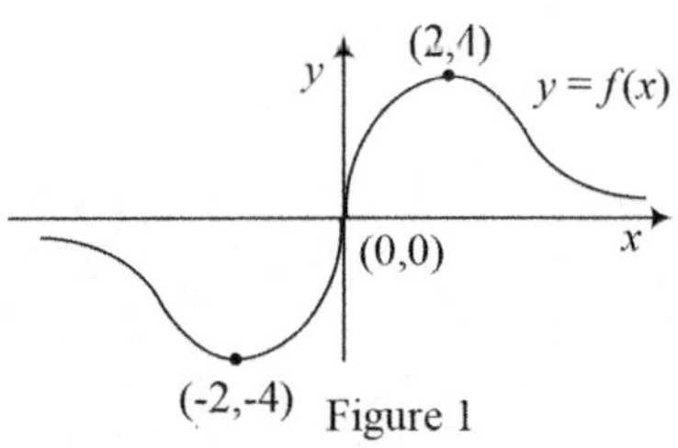

Figure 1

60. The graph of $y = f(x)$ is shown in Figure 1. Which of the following statements must be true?

I. $f(x)$ is an even function.
II. $f(x)$ is an odd function.
III. $f(-x) = f(x)$

F. I only
G. II only
H. III only
J. I and II only
K. II and III only

END OF TEST 4
STOP! DO NOT TURN THE PAGE UNTIL TOLD TO DO SO.
DO NOT RETURN TO THE PREVIOUS TEST.

Answer Key
Model Test No. 04

1	E	16	J	31	E	46	G
2	F	17	B	32	J	47	D
3	D	18	H	33	C	48	K
4	G	19	A	34	J	49	B
5	D	20	K	35	E	50	K
6	H	21	B	36	H	51	B
7	E	22	F	37	B	52	G
8	H	23	C	38	H	53	A
9	E	24	H	39	E	54	G
10	F	25	B	40	F	55	A
11	A	26	J	41	D	56	J
12	H	27	E	42	H	57	C
13	D	28	J	43	C	58	G
14	G	29	C	44	F	59	C
15	D	30	J	45	B	60	G

How to Score the ACT Mathematics Test

When you take an actual ACT Mathematics Test, your answer sheet will be "read" by a scanning machine that will record your responses to each question. Then a computer will compare your answers with the correct answers and produce your raw score. You get one point for each correct answer. For each wrong answer, there is no penalty. Therefore, you must guess and answer all the questions to get a higher score. This raw score is converted to a scaled score that is reported to you and to the colleges you specify.

Finding Your Scale Test Score

STEP 1: Count the number of right answers.
Enter the total here: ______________________

STEP 2: The number you obtained in Step 1 is your raw score.
Look up the conversion table for ACT-Math, and find your score.

Scaled Score Conversion Table ACT Mathematics Test	
Scale Score	Mathematics
36	60
35	58-59
34	56-57
33	55
32	54
31	52-53
30	50-51
29	48-49
28	46-47
27	43-45
26	41-42
25	39-40
24	37-38
23	35-36
22	33-34
21	31-32
20	30
19	27-29
18	25-26
17	22-24
16	18-21
15	15-17
14	12-14
13	09-11
12	08
11	06-07
10	05
09	04
08	-
07	03
06	02
05	-
04	-
03	01
02	-
01	00

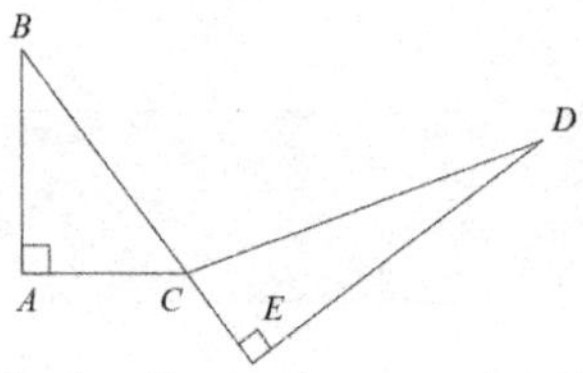

1. In the figure above, point C lies on $\overline{BE}$. If $AB = 8$, $AC = 6$, $CD = 13$, and $DE = 12$, what is the length of $\overline{BE}$?

A. 11 **B.** 12 **C.** 13 **D.** 14 **E.** 15

$\overline{BE} = \overline{BC} + \overline{CE}$, where $\overline{BC} = \sqrt{8^2 + 6^2} = \sqrt{100} = 10$ and $\overline{CE} = \sqrt{13^2 + 12^2} = \sqrt{25} = 5$

$\therefore \overline{BE} = 10 + 5 = 15$

Ans. E

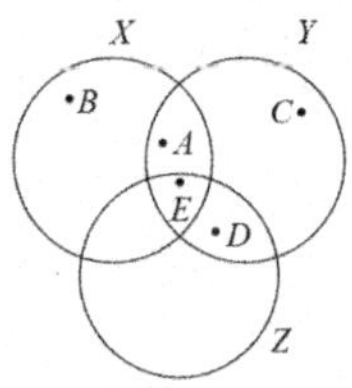

2. In the figure above, which of the labeled points is inside circle X and circle Y, but not in circle Z?

F. A **G.** B **H.** C **J.** D **K.** E

The points that X and Y share are A and E, but E is also in Z. $\therefore A$ is the point not in Z.

Ans. F

3. If $\dfrac{2a^2b}{ab^2} = \dfrac{3}{2}$, then $\dfrac{a}{b} =$

A. 3 **B.** $\dfrac{3}{2}$ **C.** $\dfrac{4}{3}$ **D.** $\dfrac{3}{4}$ **E.** $\dfrac{1}{3}$

$\dfrac{2a^2b}{ab^2} = \dfrac{2a}{b} = \dfrac{3}{2}$ $\quad \therefore \dfrac{a}{b} = \dfrac{3}{2} \cdot \dfrac{1}{2} = \dfrac{3}{4}$

Ans. D

4. The measure of the smallest angle in a certain triangle is one fifth the sum of the measures of the remaining angles. What is the measure of the smallest angle?

F. 20° **G.** 30° **H.** 36° **J.** 45° **K.** 50°

Let the sum of the remaining angles be x,

then $180 = x + \dfrac{1}{5}x = \dfrac{6}{5}x \quad \therefore x = 180 \times \dfrac{5}{6} = 150$

$\therefore$ The smallest angle $= 180 - 150 = 30$

Ans. G

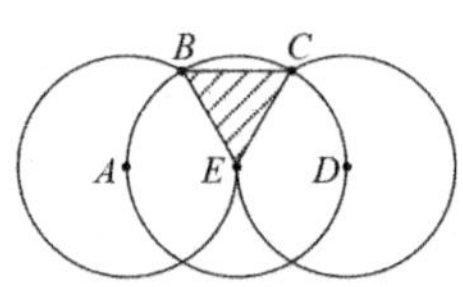

5. In the figure above, three circles, each with radius 6, have centers A, E, and D, respectively. Two circles are tangent at E, and circles intersect at B and C, as shown. What is the area of triangle EBC shown above?

A. 18 **B.** $18\sqrt{3}$ **C.** $6\sqrt{3}$ **D.** $9\sqrt{3}$ **E.** 9

First, $\overline{AB} = \overline{AE} = r$, because they are the radius of the first circle. So is, $\overline{BE} = \overline{CE} = r$ and $\overline{ED} = \overline{CD} = r$ for the second and third circles. $\therefore \overline{AB} = \overline{AE} = \overline{BE} = \overline{CE} = \overline{ED} = \overline{CD} = r = 6$

Therefore, ΔEBC is an equilateral triangle with side length 6.

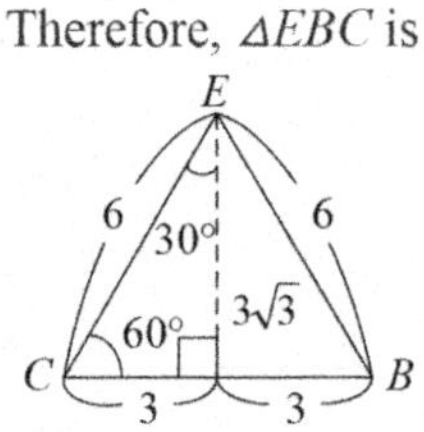

$\therefore$ The area $= \dfrac{1}{2}(6)(3\sqrt{3}) = 9\sqrt{3}$

Ans. D

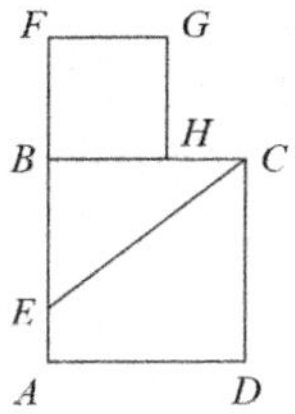

6. In the figure above, $ABCD$ and $BFGH$ are squares, $4BE = 3AB$ and $FE = CE$. If $BF = 1$, what is the length of $\overline{AD}$?

F. $\frac{2}{3}$

G. $\frac{3}{2}$

H. 2

J. $\frac{5}{2}$

K. 3

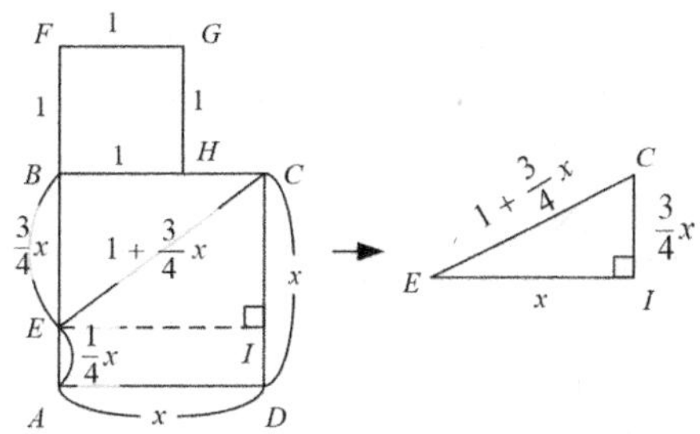

Referring to the above figure, since $4BE = 3AB = 3x$,

we get $BE = \frac{3}{4}AB = \frac{3}{4}x$. Also, $FE = CE$. $\therefore 1 + \frac{3}{4}x = CE$

Now, using Pythagorean Th., $(\frac{3}{4}x)^2 + x^2 = (1 + \frac{3}{4}x)^2$

$\therefore (\frac{3}{4}x)^2 + x^2 = 1 + 2\frac{3}{4}x + (\frac{3}{4}x)^2$

$\therefore x^2 - \frac{3}{2}x - 1 = 0$, or $2x^2 - 3x - 2 = 0$

$\therefore (2x + 1)(x - 2) = 0$ $\therefore x = 2, x \neq -\frac{1}{2}$

Ans. H

7. If $5 - 3x = 5$, what is the value of $(5 - 3x)(5 + 3x)$?

A. 5 **B.** 10 **C.** 15 **D.** 20 **E.** 25

$5 - 3x = 5$ $\therefore 5 - 5 = 3x$ $\therefore x = 0$

$(5 - 3x)(5 + 3x) = (5 - 0)(5 + 0) = 25$

Ans. E

8. The measures of the three angles of a triangle are $5a°$, $3a°$and $2a°$. The measure of the largest angle of the triangle is how much greater than the measure of the smallest angle?

F. 15° **G.** 36° **H.** 54° **J.** 72° **K.** 105°

$5a + 3a + 2a = 10a = 180°$ $\therefore a = 18°$

$\therefore$ The smallest angle $= 2a = 36°$,

and the largest angle $= 5a = 90°$ $\therefore 90° - 36° = 54°$

Ans. H

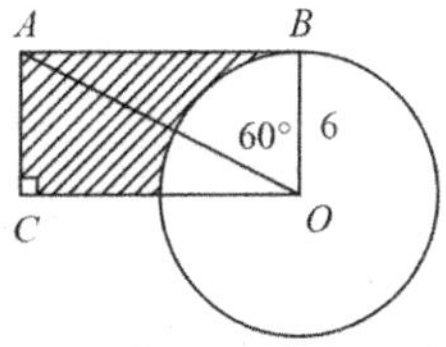

9. In the figure above, the center of the circle is O and $\overline{AB}$ is tangent to the circle at point B. What is the area of the shaded region of the rectangle $ABOC$?

A. $36\sqrt{3} - 6\pi$ **B.** $36\sqrt{3} - 12\pi$ **C.** $18\sqrt{3} - 6\pi$

D. $18\sqrt{3} - 12\pi$ **E.** $36\sqrt{3} - 9\pi$

The area of □$ABOC$

$= 6\sqrt{3} \times 6 = 36\sqrt{3}$.

Also, the area of $\frac{1}{4}$ circle

$= \frac{1}{4} \times \pi \times (6)^2 = 9\pi$

$\therefore$ Area $= 36\sqrt{3} - 9\pi$

Ans. E

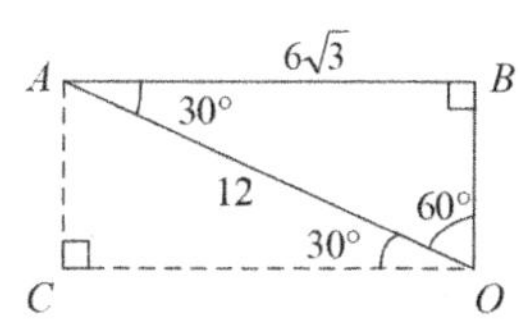

10. In the xy-plane, line l passes through the point (1, 2) and the vertex of the parabola with the equation $y = 2x^2 - 12x + 12$. What is the slope of line l?

F. - 4 **G.** $-\frac{1}{4}$ **H.** 0 **J.** $\frac{1}{4}$ **K.** 4

$y = 2x^2 - 12x + 12 = 2(x^2 - 6x + 6) = 2[(x-3)^2 - 9 + 6]$
$= 2(x-3)^2 - 6$ ∴ Vertex, $V = (3, -6)$
Therefore, the slope of line l between (3, -6) and (1, 2) is
$M = \frac{-6-2}{3-1} = -\frac{8}{2} = -4$

Ans. F

NUMBER OF YEARS OF DRIVING EXPERIENCE

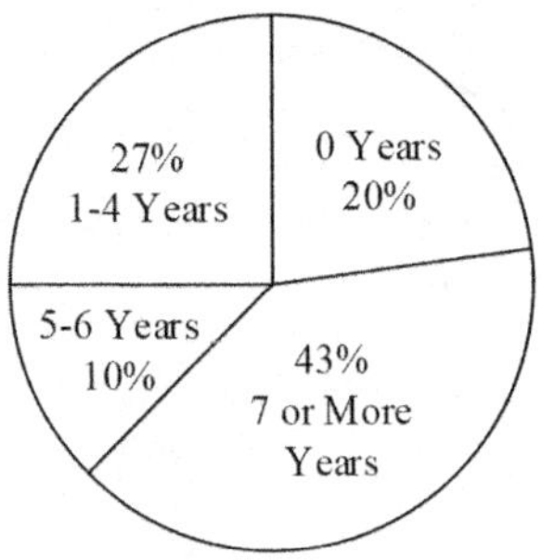

11. The graph above shows the distribution of the number of years of driving experience for a group of 1,000 adults. According to the graph, how many of these adults had 5 or more years of driving experience?

A. 530 **B.** 620 **C.** 740 **D.** 180 **E.** 390

$1{,}000 \times 0.53 = 530$

Ans. A

12. If 3.38 is rounded to the nearest tenth and the result is tripled, what is the final result?

F. 10.4 **G.** 10.3 **H.** 10.2 **J.** 10.1 **K.** 10.0

$3.4 \times 3 = 10.2$

Ans. H

13. At a book sale, all paperback books cost d dollars each and all hardback books cost $3d$ dollars each. If Emily paid $13d$ dollars for 7 books at this sale, how many hardback books did she buy?

A. None **B.** One **C.** Two **D.** Three **E.** Four

Let p = number of paperback books and
h = number of hardback books.
Then, $p + h = 7$ - eq ①
$13d = (d \times p) + (3d \times h)$ - eq ②
Now, $p = 7 - h$ into eq ② gives $13d = d \times (7 - h) + 3dh$,
or $13 = 7 - h + 3h$
∴ $h = 3$

Ans. D

14. For which of the following functions is it true that $f(-x) = f(x)$ for all values of x?

F. $f(x) = x + 4$ **G.** $f(x) = x^2 + 4$ **H.** $f(x) = x^3 + 4$
J. $f(x) = x^2 + x$ **K.** $f(x) = x^3 + x$

$f(-x) = f(x)$
∴ It is even-function, or it is symmetric to y-axis.
Using graphing utility, $f(x) = x^2 + 4$ is the only even function.

Ans. G

15. Which of the following represents half the difference between $4x$ and 14?

A. $7 - 4x$ **B.** $3x - 7$ **C.** $2x - 14$
D. $2x - 7$ **E.** $14 - 2x$

$\frac{1}{2}(4x - 14) = 2x - 7$

Ans. D

GROWTH OF A POPULATION OF BACTERIA

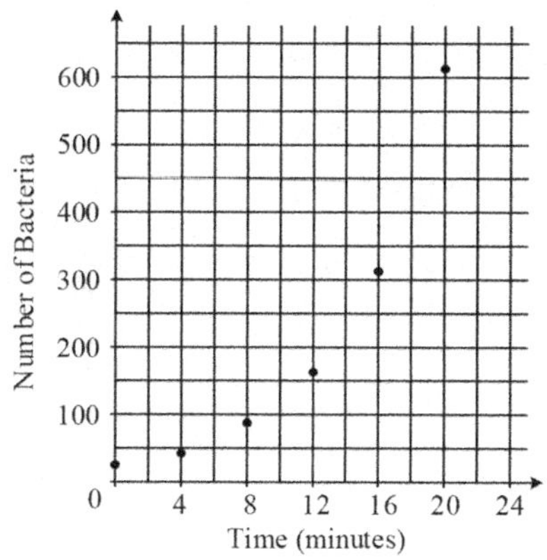

16. The number of bacteria in a dish at various times from the start of an experiment is shown in the graph above. Based on the pattern indicated by the graph, which of the following is closest to the number of minutes from the start of the experiment to the time when there were 450 bacteria in the dish?

F. 6 **G.** 10 **H.** 14 **J.** 18 **K.** 22

At $t = 16$ min, we have approximately 320 bacteria, while at $t = 20$ min, we have about 620 bacteria.
$\therefore$ 450 bacteria happens approximately at $t = 18$ min.

Ans. J

17. If j and k are integers and $0 \le j \le k \le 6$, which of the following is not a possible value of jk?

A. 9 **B.** 14 **C.** 16 **D.** 18 **E.** 20

The possible values of j and k are, 0, 1, 2, 3, 4, 5, 6.
For the product of $j \times k$, 14 is not possible, because 14 is a product of prime number 2×7, which we do not have 7 in our choices.

Ans. B

18. If the average (arithmetic mean) of x, y and 12 is 8, what is the average of x and y?

F. 2
G. 3
H. 6
J. 7
K. It cannot be determined from the information given

$\frac{x+y+12}{3} = 8$
$\therefore x + y + 12 = 3 \times 8 = 24$
$\therefore x + y = 12$
$\therefore$ Avg. of $(x + y) = \frac{12}{2} = 6$

Ans. H

19. If $2 + \sqrt{x} = 4$, then which of the following is $(x + 2)^2$?

A. 36 **B.** 24 **C.** 18 **D.** 12 **E.** 9

$2 + \sqrt{x} = 4 \therefore \sqrt{x} = 2 \therefore x = 4$
$\therefore (x + 2)^2 = (4 + 2)^2 = 6^2 = 36$

Ans. A

20. If $t = 2x - 3$ and $t = y - 7$, which of the following must be true?

F. $t = x$
G. $x = y$
H. $x < y$
J. $x > y$
K. infinitely many solutions

$t = 2x - 3 = y - 7$, or $2x - 3 = y - 7$.
Here, there are infinitely many solutions.

Ans. K

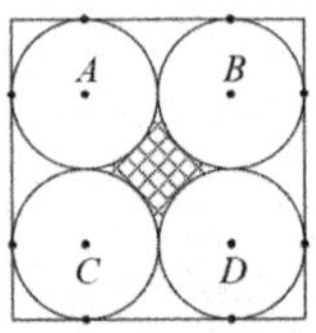

21. In the figure above, each of the points A, B, C and D is the center of a circle of radius 1. The points where the circles are tangent to each other and to the square are marked. What is the area of the shaded part?

A. $8 - \pi$ **B.** $4 - \pi$ **C.** π **D.** 4π **E.** 2π

The area of the shaded part becomes,

$(2 \times 2) - 4(\frac{1}{4}\pi r^2) = 4 - \pi(1)^2 = 4 - \pi$

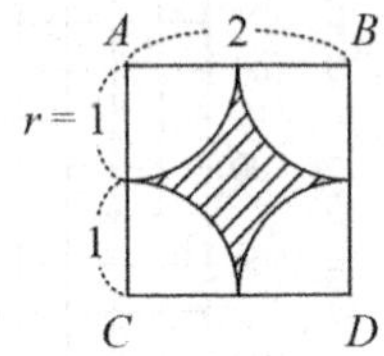

Ans. B

22. If n is a positive integer, which of the following is equivalent to $9^n + 9^n$?

F. $2 \cdot 3^{2n}$ **G.** 3^{4n} **H.** $(2 \cdot 3 \cdot 3)^n$
J. $(2 \cdot 3)^{2n}$ **K.** $(3 \cdot 3 \cdot 3 \cdot 3)^{2n}$

$9^n + 9^n = (3^2)^n + (3^2)^n = 2(3^2)^n = 2 \times 3^{2n}$

Ans. F

23. If a number is chosen at random from the set {-8, -6, -4, -2, 0, 2, 4, 6, 8}, what is the probability that it is a member of the solution set of both $|x - 1| < 7$ and $|2x + 5| > 3$?

A. 0 **B.** $\frac{2}{3}$ **C.** $\frac{4}{9}$ **D.** $\frac{5}{9}$ **E.** $\frac{7}{9}$

First, $|x - 1| < 7$, $\therefore -7 < x - 1 < 7$ $\therefore -6 < x < 8$
Also, $|2x + 5| > 3$, $\therefore 2x + 5 > 3$ $\therefore x > -1$ or $2x + 5 < -3$, $\therefore x < -4$ Therefore, we have

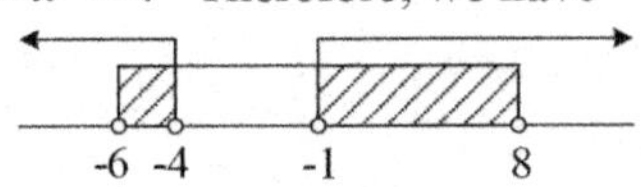

The numbers which satisfy the above condition are, -5, 0, 1, 2, 3, 4, 5, 6 and 7. Among these, only 0, 2, 4 and 6 are in the set of 9 numbers. $\therefore P = \frac{4}{9}$

Ans. C

24. $(2^x \cdot 2^{3x})^{2x} =$

F. 2^{6x^3} **G.** 2^{6x} **H.** 2^{8x^2} **J.** 2^{8x} **K.** 2^{16x^2}

$(2^{x+3x})^{2x} = (2^{4x})^{2x} = 2^{8x^2}$

Ans. H

25. The eight letters a, b, c, d, e, f, g and h represent the eight integers 2 through 9, not necessarily in that order. If b is the product of prime numbers e and h, which of the following could NOT be the value of f?

A. 5 **B.** 6 **C.** 7 **D.** 8 **E.** 9

Among the eight integers 2 through 9, there are only 2, 3, 5 and 7 as prime numbers. Therefore, if b = prime × prime, then the only choice for b is 6, where $b = 6 = 2 \times 3 = e \times h$
$\therefore b = 6$, $e = 2$, $h = 3$ Therefore, f cannot be 6.

Ans. B

26. The radius of a wheel is 6 feet. If the wheel rolls 5 miles without slipping, how many revolutions will it make? (1 mile = 5,280 feet)

F. 400 **G.** 500 **H.** 600 **J.** 700 **K.** 900

$c = 2\pi r = 2\pi(6) = 12\pi$ ft. = 1 revolution
Also, 5 mile = 5 × 5,280 ft. = 26,400 ft.
$\therefore 26{,}400 \div (12 \times 3.14) = 700.64$

Ans. J

27. If $a - 3b$ is 5 more than $2a - 2b$, the value of which of the following expressions can be determined?

A. a **B.** b **C.** $a - 3b$ **D.** $a - b$ **E.** $a + b$

$a - 3b = 5 + (2a - 2b)$
$\therefore a - 3b - 2a + 2b = -a - b = 5$ $\therefore a + b = -5$

Ans. E

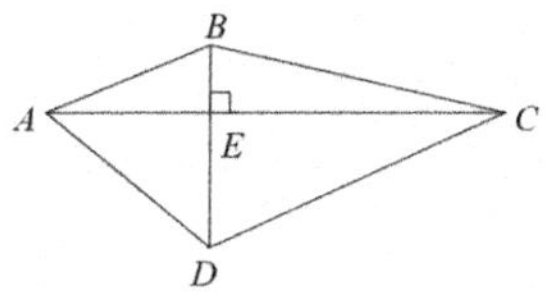

28. In $ABCD$ above, $AC = 15$, $BD = 8$ and $\overline{AC} \perp \overline{BD}$. What is the area of quadrilateral $ABCD$?

F. 15 **G.** 30 **H.** 45 **J.** 60 **K.** 120

Area $= \frac{1}{2}(\text{Diag. } \overline{AC} \times \text{Diag. } \overline{BD}) = \frac{1}{2}(15 \times 8) = 60$

Ans. J

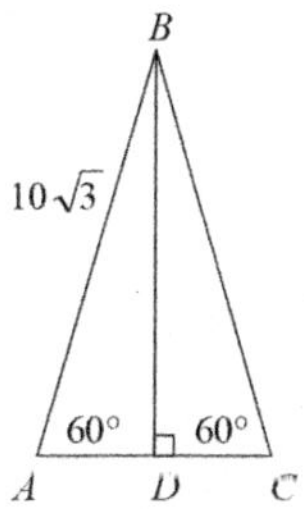

29. In $\triangle ABC$ above, what is the length of BD?

A. $7\sqrt{3}$
B. $8\sqrt{3}$
C. 15
D. 17
E. $11\sqrt{3}$

$\therefore BD = 15$

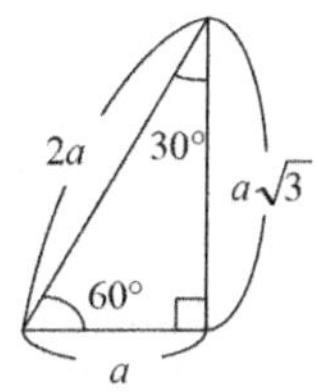

∴

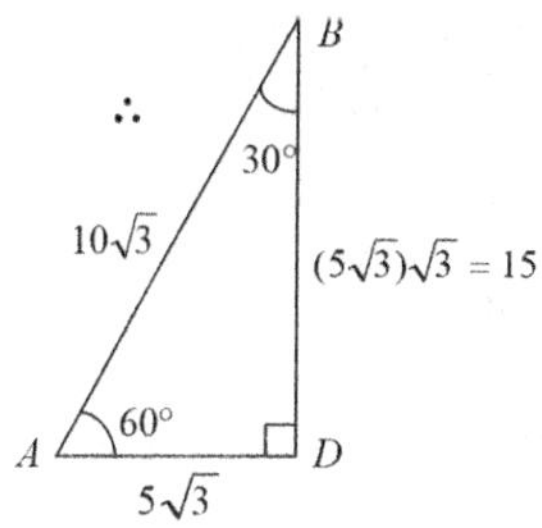

Ans. C

30. If $2^{2n+1} = 8^5$, what is the value of n?

F. 3 **G.** 4 **H.** 5 **J.** 7 **K.** 8

$2^{2n+1} = (2^3)^5 = 2^{15}$

$\therefore 2n + 1 = 15 \quad \therefore 2n = 14 \quad \therefore n = 7$

Ans. J

31. In a group of 500 students, 350 are studying Spanish, 108 are studying Chinese, and 26 are studying Latin. What is the least possible number of these students that might NOT be studying any of these languages?

A. 10 **B.** 12 **C.** 13 **D.** 14 **E.** 16

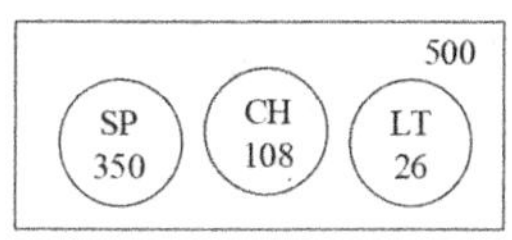

$\therefore$ Least $= 500 - 350 - 108 - 26 = 16$

Ans. E

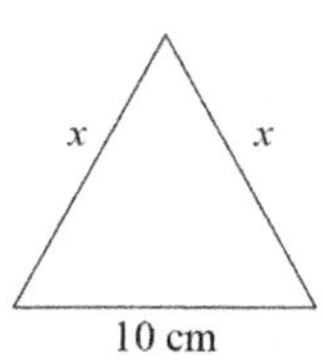

32. If the perimeter of the triangle above is 36 centimeters, what is its area, in square centimeters?

F. 48 **G.** 54 **H.** 58 **J.** 60 **K.** 64

$\therefore$ area $= \frac{1}{2}(10)(12) = 60$

(Refer to special right triangle, 5:12:13 !!)

Ans. J

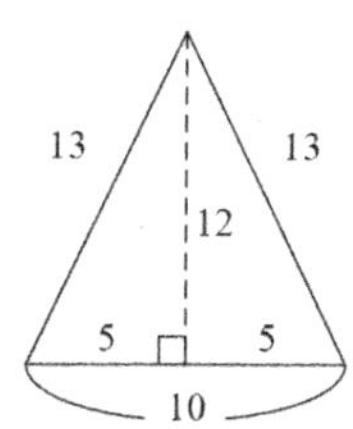

33. What is the number that satisfies the following three conditions?

- It is 4 digit integer less than 1200.
- The sum of its digits is 12.
- Its tens and units digits are the same.

A. 1074 **B.** 1065 **C.** 1155 **D.** 1164 **E.** 1173

Since it is 4 digit and less than 1,200,
we get the first 2 digits as, [1][1][1][1]. Also,
the last 2 digit are the same with sum of all digits being 12.
∴ [1][1][5][5].

Ans. C

34. Point C is on line segment AB such that $\frac{AC}{AB} = \frac{3}{5}$. If the length of AC is 9, what is the length of CB?

F. 4.8 **G.** 5 **H.** 5.2 **J.** 6 **K.** 6.4

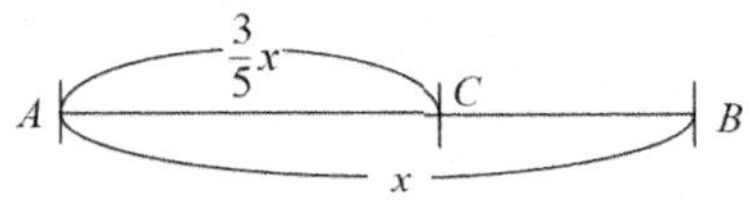

Now, $\overline{AC} = 9 = \frac{3}{5}x$

$\therefore x = 15 \quad \therefore \overline{CB} = 15 - 9 - 6$

Ans. J

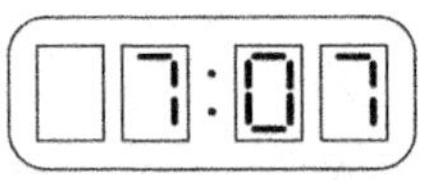

35. The face of a 12-hour digital clock, pictured above, shows one example of a time at which reading the digits from left to right is the same as reading the digits from right to left. How many such times would be shown on this clock in a 12-hour period starting at 12 midnight?

A. 49 **B.** 52 **C.** 54 **D.** 55 **E.** 57

Let's start with one o'clock.

Then, 1:01, 1:11, 1:21, 1:31, 1:41, 1:51
2:02, 2:12, 2:22, 2:32, 2:42, 2:52
…
9:09, 9:19, 9:29, 9:39, 9:49, 9:59
Each with 6 possible cases.

$\therefore 9 \times 6 = 54$ and then, 10:01, 11:11, 12:21, 3 possible cases.

$\therefore$ Total $= 54 + 3 = 57$

Ans. E

36. The first term of a sequence of numbers is 12. If each term after the first is 4 more than 3 times the preceding term, what is the fourth term of this sequence?

F. 364 **G.** 372 **H.** 376 **J.** 384 **K.** 396

$\underline{12}$, $\underline{4 + 3(12) = 40}$, $\underline{4 + 3(40) = 124}$, $\underline{4 + 3(124) = 376}$

Ans. H

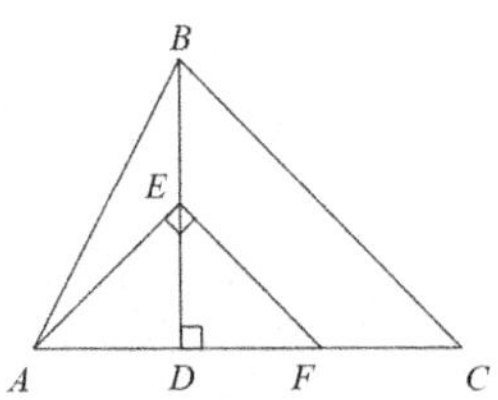

37. In $\triangle ABC$ above, $AD = ED$, $BC = 2EF$ and $\overline{BC} \parallel \overline{EF}$. What is the value of the fraction $\frac{\text{area } \square AEF}{\text{area } \square ABC}$?

A. $\frac{2}{9}$ **B.** $\frac{1}{3}$ **C.** $\frac{4}{9}$ **D.** $\frac{1}{2}$ **E.** $\frac{5}{9}$

Since $\triangle DEF$ is similar triangle to $\triangle DBC$ ($\because BC \parallel EF$, and $\overline{BC} = 2\overline{EF}$) with $\frac{1}{2}$ ratio, we may say

$\overline{AD} = \overline{DF} = \overline{FC} = \overline{BE} = x$.

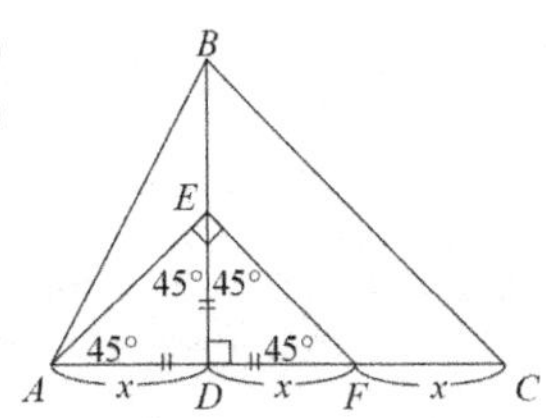

$\therefore$ Area of $\triangle AEF = \frac{1}{2}(2x)(x) = x^2$,

while area of $\triangle ABC = \frac{1}{2}(3x)(2x) = 3x^2 \quad \therefore \frac{1}{3}$

Ans. B

38. If $-1.7 < \frac{n}{7} < -1.2$, where n is an integer, what is one possible value of $|n|$?

F. 7 **G.** 8 **H.** 10 **J.** 12 **K.** 15

$(-1.7)(7) < n < (-1.2)(7)$

$\therefore -11.9 < n < -8.40 \quad \therefore |n| = |-11|, |-10|$ or $|-9|$

$\therefore 9, 10$ or 11

Ans. H

39. Ali, Ben, Chris, David and Emily are each to be assigned to 1 of 2 teams. How many different assignments of these 5 people are possible if at least 1 person must be assigned to each team?

A. 5
B. 10
C. 15
D. 20
E. 30

Team 1 Team 2

For Team 1,

One of 5 people is possible. $\therefore {}_5C_1 = 5$

Two of 5 people is possible. $\therefore {}_5C_2 = 10$

Three of 5 people is possible. $\therefore {}_5C_3 = 10$

Four of 5 people is possible. $\therefore {}_5C_4 = 5$

But five of five people is NOT possible, because at least one must be assigned to Team 2!!

$\therefore 5 + 10 + 10 + 5 = 30$

Ans. E

40. If p, r, and s are any three consecutive positive integers, which of the following must be even?

I. prs
II. $p + r + s$
III. $p^2 + r^2 + s^2$

F. I only **G.** II only **H.** III only
J. I and III only **K.** I, II, and III

Let's try both of the following cases:

$p = 1, r = 2, s = 3$ and $p = 2, r = 3, s = 4$

The only choice that works for both 1) and 2) to be even number is (I).

Ans. F

41. If x is a percent of b and if b is c percent of y, then x is what percent of y? (Assume all variables are positive.)

A. $\frac{a+b+c}{100}$% **B.** $\frac{ab}{100}$% **C.** $\frac{bc}{100}$% **D.** $\frac{ac}{100}$% **E.** $\frac{abc}{100}$%

$x = a \times \frac{1}{100} \times b$ and $b = c \times \frac{1}{100} \times y$.

By replacing b, $x = \frac{a}{100} \times (c \times \frac{1}{100} \times y)$, or $x = \frac{ac}{100} \cdot \frac{1}{100} \times y$

$\therefore x$ is $\frac{ac}{100}$ percent of y.

Ans. D

42. When the number x is multiplied by 7, the result is the same as when 2 is added to x. What is the value of $4x$?

F. $\frac{3}{4}$ **G.** 1 **H.** $\frac{4}{3}$ **J.** 3 **K.** 4

$7x = 2 + x$

$\therefore 6x = 2 \quad \therefore x = \frac{1}{3} \quad \therefore 4x = 4\frac{1}{3} = \frac{4}{3}$

Ans. H

5.10100100001000000001...

43. The decimal number above consists of only 1's and 0's to the right of the decimal point. The first 1 is followed by one 0, the second 1 is followed by two 0's, the third 1 is followed by four 0's, the fourth 1 is followed by eight 0's and so on. What is the total number of 0's between the 9th and the 11th 1 in this decimal number?

A. 256 **B.** 512 **C.** 768 **D.** 1536 **E.** 1792

For this problem, let's look at the following patterns:

The first 1, followed by one 0's, $\therefore 1 = 2^0$

The second 1, followed by two 0's, $\therefore 2 = 2^1$

The third 1, followed by four 0's, $\therefore 4 = 2^2$

The fourth 1, followed by eight 0's, $\therefore 8 = 2^3$

...

The 9th 1, followed by (2^{9-1}) 0's,

The 10th 1, followed by (2^{10-1}) 0's.

Therefore, $2^8 + 2^9 = 768$.

Ans. C

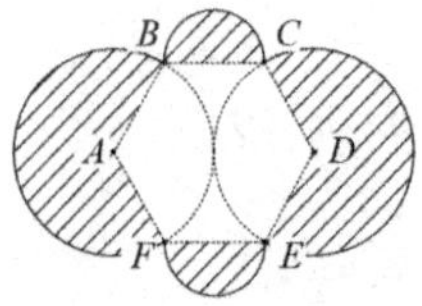

44. In the figure above, $ABCDEF$ is a regular hexagon whose sides are each 6 units long. A and D are the centers of the two large circles, and BC and FE are diameters of the two small semicircles. What is the total area of the figure?

F. 57π **G.** 54π **H.** 55π **J.** 56π **K.** 50π

Since ABCDEF is a regular hexagon,
$\overline{AB} = \overline{AF} = \overline{FE} = \overline{ED} = \overline{DC} = \overline{BC} = r = 6$
Also, $\angle BAF = 120°$ $\therefore$ Total area of the shaded part is,
$[\frac{2}{3}(\pi(6)^2) \times 2] + [\frac{1}{2}(\pi(3)^2) \times 2] = 48\pi + 9\pi = 57\pi$

Ans. F

45. There are only white and black marbles in a box. If 4 marbles are white and the probability of choosing a white marble at random from the box is $\frac{1}{4}$, how many black marbles are in the box?

A. 18 **B.** 12 **C.** 9 **D.** 6 **E.** 2

P (choosing white) $= \frac{4}{T} = \frac{1}{4}$
$\therefore$ Total T = 16 $\therefore$ Black = 16 − 4 = 12

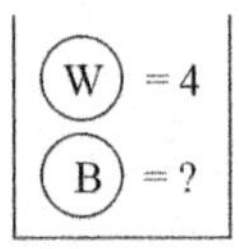

Ans. B

46. If n and k are positive odd integers, which of the following has the least value?

F. n^k **G.** $(-n)^{-k}$ **H.** $(-n)^{2k}$ **J.** $(-2n)^{-k}$ **K.** $(-2n)^{-2k}$

Let's try with $n = 3$ and $k = 1$. Then, the least value will be, **G.**
$(-3)^{-1} = -\frac{1}{3}$

Ans. G

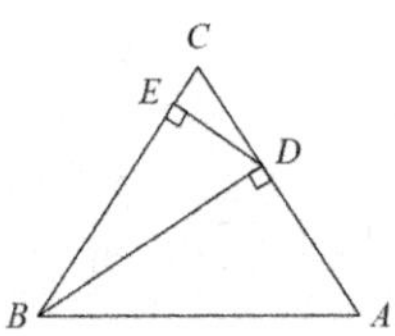

47. In the figure above, $\triangle ABC$ is an equilateral triangle. What is the ratio of the areas of $\triangle ABC$ to $\triangle CDE$?

A. 2:1
B. 4:1
C. 6:1
D. 8:1
E. 16:1

Refer to the above equilateral $\triangle$,
let's say each side $x = 4$.
Then, since $\overline{BD}$ is a perpendicular bisector of $\overline{AC}$, $\overline{CD} = \overline{DA} = 2$.
Also, $\triangle ABD$ is a right $\triangle$ with 30°–60°–90°. $\therefore \overline{BD} = 2\sqrt{3}$

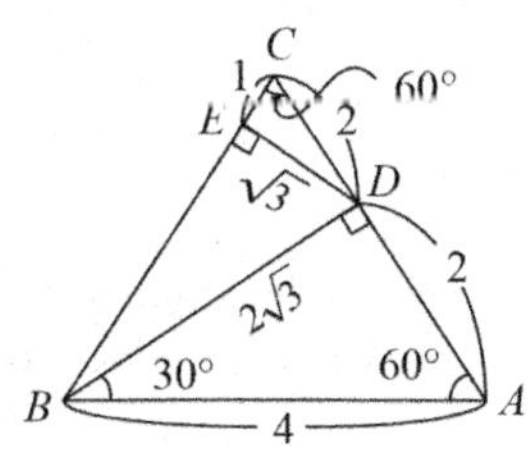

$\therefore$ area of $\triangle ABC = \frac{1}{2}(4)(2\sqrt{3}) = 4\sqrt{3}$

Also, area of $\triangle CDE = \frac{1}{2}(1)\sqrt{3} = \frac{\sqrt{3}}{2}$

$\therefore \triangle ABC : \triangle CDE = 4\sqrt{3} : \frac{\sqrt{3}}{2} = 8 : 1$

Ans. D

48. If a wheel revolves at the rate of $\frac{2}{3}\pi$ per second, how many complete revolutions does it make in one <u>minute</u>?

F. 9,000 **G.** 150 **H.** 144 **J.** 40 **K.** 20

Let the angular speed, $W = \frac{2\pi}{3}$/sec. Then, for one minute,
the total angle $T = \frac{2\pi}{3} \times 60 = 40\pi = 40(180°) = 7200°$.
But one revolution is 360°. $\therefore$ 7200 ÷ 360 = 20 rev.

Ans. K

49. The least integer of a set of consecutive integers is -25. If the sum of these integers is 26, what is the average of these integers?

A. 0 **B.** $\frac{1}{2}$ **C.** $\frac{1}{4}$ **D.** 1 **E.** $\frac{26}{51}$

The sum of all becomes,
$-25 - 24 - 23 \ldots 3 - 2 - 1 + 0 + 1 + 2 + 3 \ldots 23 + 24 + 25 = 26$
From (-)25 to (+)26, there are $26 - (-25) + 1 = 52$ integers.
$\therefore$ Avg. $= \frac{26}{52} = \frac{1}{2}$

Ans. B

50. If a and b are positive integers, which of the following expressions is equivalent to $\frac{(2^a \cdot 2^a)^b}{2^a}$?

F. 2^{a+b} **G.** 2^{ab} **H.** 2^{2b} **J.** $2^{a(ab-1)}$ **K.** $2^{a(2b-1)}$

$\frac{(2^a \cdot 2^a)^b}{2^a} = \frac{(2^{2a})^b}{2^a} = 2^{2ab-a} = 2^{a(2b-1)}$

Ans. K

51. If a person needs 3 feet of space along the edge of a circular table to be seated comfortably, what is the maximum number of people who can be seated comfortably at a table of radius 7 feet?

A. 12 **B.** 14 **C.** 7 **D.** 21 **E.** 15

$C = 2\pi(7) = 14\pi = 43.98$
$\therefore 43.98 \div 3 = 14.66$
$\therefore$ 14 people

Ans. B

52. The sum of 5 consecutive even integers is 1,500. What is the least of these integers?

F. 326 **G.** 296 **H.** 272 **J.** 300 **K.** 216

Let $2n$ be the first even consecutive number.
Then, $2n + (2n + 2) + (2n + 4) + (2n + 6) + (2n + 8)$
$= 10n + 20 = 1500$
$\therefore 10n = 1480 \quad \therefore n = 148 \quad \therefore 2n = 2(148) = 296$

Ans. G

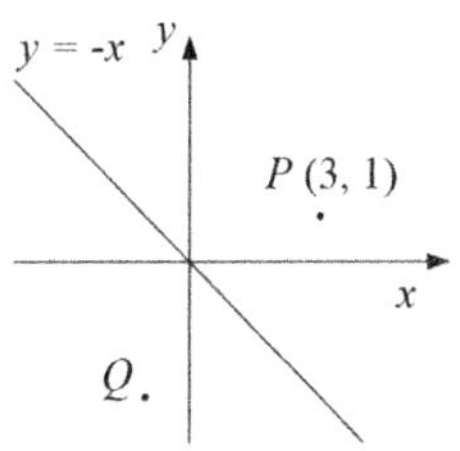

53. In the figure above, point P is the reflection of point Q through the line $y = -x$. What is the coordinates of point Q?

A. (-1, -3) **B.** (-1, -2) **C.** (-3, -1)
D. $(-\frac{3}{2}, -\frac{1}{2})$ **E.** (-2, -1)

For this reflection on $y = -x$, we need to switch x into y, and y into x, and then negative both x and y points of P.
$\therefore P(3, 1) \rightarrow Q(-1, -3)$

Ans. A

54. Peter slices a pizza into k equal pieces and eats three pieces. In terms of k, what percent of the pizza is left?

F. $100(k - 3)$ % **G.** $\frac{100(k-3)}{k}$ %

H. $\frac{100k}{k-3}$ % **J.** $\frac{k-3}{100}$ %

K. $\frac{k-3}{100k}$ %

$\frac{k-3}{k} \times 100\%$

Ans. G

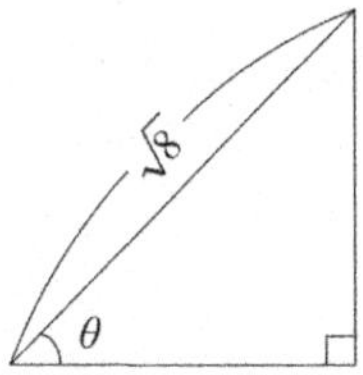

55. In Figure 1, if $\theta = 45°$, what is the area of the triangle?

A. 2 **B.** 2.36 **C.** 4 **D.** 4.72 **E.** 6

When $\theta = 45°$, we get 45°-45°-90° triangle with 1-1-$\sqrt{2}$ ratio.
Since $a\sqrt{2} = \sqrt{8} = 2\sqrt{2}$, we have $a = 2$.
Therefore, the area is, $\frac{1}{2}(2)(2) = 2$

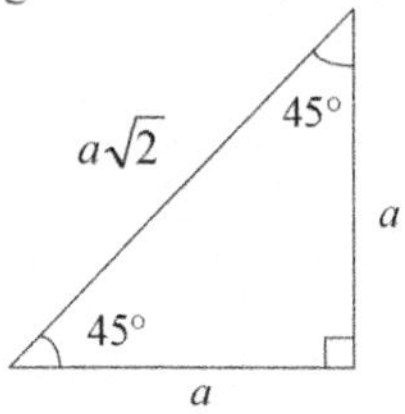

Ans. A

56. If $\sqrt{(x+2)} = 3$, then $x =$

F. -3.5 **G.** 2 **H.** 4.2 **J.** 7 **K.** 12

Squaring both sides, we get $x + 2 = 3^2 = 9$
Therefore, $x = 9 - 2 = 7$

Ans. J.

57. If $f(7x + 4) = 28x + 16$ for all real x, what is $f(x)$?

A. $x + 5$ **B.** $7x + 5$ **C.** $4x$
D. $2x + 5$ **E.** $2x - 5$

$f(7x + 4) = 28x + 16 = 4(7x + 4)$
Therefore, if we let $u = 7x + 4$,
then $f(u) = 4u$ or $f(x) = 4x$ is the answer.

Ans. C

58. If the line $y = mx + b$ intersects the parabola $y = 2x^2 + 3$ at $x = 0$, what is the y-intercept of the line?

F. 8 **G.** 3 **H.** 5 **J.** 4 **K.** 2.5

Since they are intersecting each other, we set, $mx + b = 2x^2 + 3$.
Also, they are intersecting at $x = 0$, we get, $m(0) + b = 2(0)^2 + 3$
$\therefore b = 3$, which is y-intercept.

Ans. G

59. $f(x) = \begin{cases} x^2 - 4 \text{ if } x > 2 \\ -2x + 4 \text{ if } x \leq 2 \end{cases}$. What is the range of f?

A. $y \leq 1$
B. $y > 1$
C. $y \geq 0$
D. $y \geq 3$
E. All real numbers

The graph of this piecewise function is,

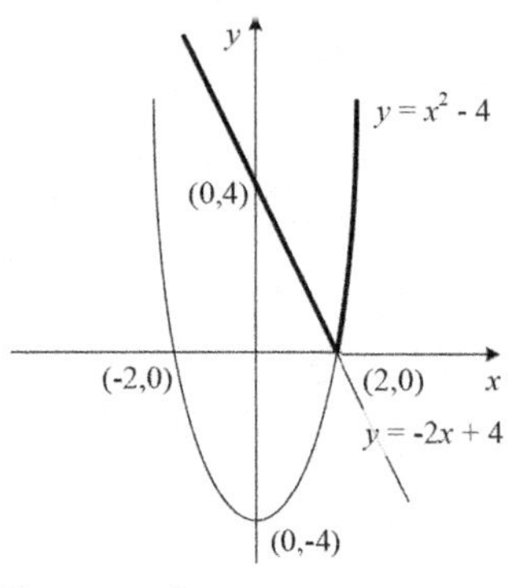

The range is $y \geq 0$.

Ans. C

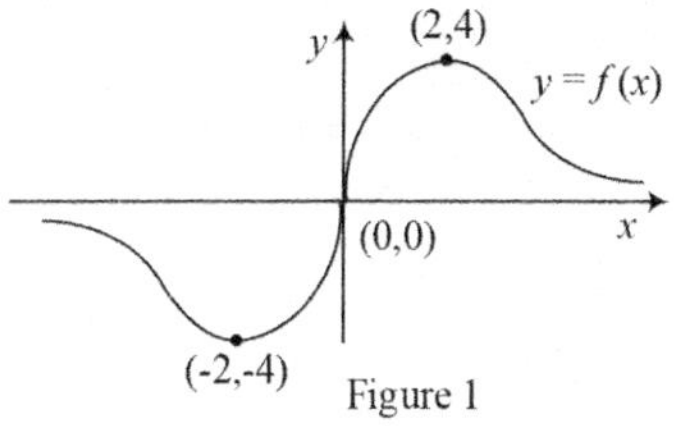

Figure 1

60. The graph of $y = f(x)$ is shown in Figure 1. Which of the following statements must be true?

I. $f(x)$ is an even function.
II. $f(x)$ is an odd function.
III. $f(-x) = f(x)$

F. I only
G. II only
H. III only
J. I and II only
K. II and III only

Since $y = f(x)$ is symmetric to origin $(0,0)$, it is odd function.

Ans. G

Model Test No. 05

60 Questions / 60 Minutes

Directions: Solve each problem, choose the correct answer, and then fill in the corresponding oval on your answer document.

Do not linger over problems that take too much time. Solve as many as you can; then return to the others in the time you have left for this test.

You are permitted to use a calculator on this test. You many use your calculator for any problems you choose, but some of the problems may best be done without using a calculator.

Note: Unless otherwise stated, all of the following should be assumed.

1. Illustrative figures are NOT necessarily drawn to scale.
2. Geometric figures lie in a plane.
3. The word line indicates a straight line.
4. The word average indicates arithmetic mean.

Answer Sheet
Model Test No. 05

1	ⒶⒷⒸⒹⒺ	16	ⒻⒼⒽⒿⓀ	31	ⒶⒷⒸⒹⒺ	46	ⒻⒼⒽⒿⓀ
2	ⒻⒼⒽⒿⓀ	17	ⒶⒷⒸⒹⒺ	32	ⒻⒼⒽⒿⓀ	47	ⒶⒷⒸⒹⒺ
3	ⒶⒷⒸⒹⒺ	18	ⒻⒼⒽⒿⓀ	33	ⒶⒷⒸⒹⒺ	48	ⒻⒼⒽⒿⓀ
4	ⒻⒼⒽⒿⓀ	19	ⒶⒷⒸⒹⒺ	34	ⒻⒼⒽⒿⓀ	49	ⒶⒷⒸⒹⒺ
5	ⒶⒷⒸⒹⒺ	20	ⒻⒼⒽⒿⓀ	35	ⒶⒷⒸⒹⒺ	50	ⒻⒼⒽⒿⓀ
6	ⒻⒼⒽⒿⓀ	21	ⒶⒷⒸⒹⒺ	36	ⒻⒼⒽⒿⓀ	51	ⒶⒷⒸⒹⒺ
7	ⒶⒷⒸⒹⒺ	22	ⒻⒼⒽⒿⓀ	37	ⒶⒷⒸⒹⒺ	52	ⒻⒼⒽⒿⓀ
8	ⒻⒼⒽⒿⓀ	23	ⒶⒷⒸⒹⒺ	38	ⒻⒼⒽⒿⓀ	53	ⒶⒷⒸⒹⒺ
9	ⒶⒷⒸⒹⒺ	24	ⒻⒼⒽⒿⓀ	39	ⒶⒷⒸⒹⒺ	54	ⒻⒼⒽⒿⓀ
10	ⒻⒼⒽⒿⓀ	25	ⒶⒷⒸⒹⒺ	40	ⒻⒼⒽⒿⓀ	55	ⒶⒷⒸⒹⒺ
11	ⒶⒷⒸⒹⒺ	26	ⒻⒼⒽⒿⓀ	41	ⒶⒷⒸⒹⒺ	56	ⒻⒼⒽⒿⓀ
12	ⒻⒼⒽⒿⓀ	27	ⒶⒷⒸⒹⒺ	42	ⒻⒼⒽⒿⓀ	57	ⒶⒷⒸⒹⒺ
13	ⒶⒷⒸⒹⒺ	28	ⒻⒼⒽⒿⓀ	43	ⒶⒷⒸⒹⒺ	58	ⒻⒼⒽⒿⓀ
14	ⒻⒼⒽⒿⓀ	29	ⒶⒷⒸⒹⒺ	44	ⒻⒼⒽⒿⓀ	59	ⒶⒷⒸⒹⒺ
15	ⒶⒷⒸⒹⒺ	30	ⒻⒼⒽⒿⓀ	45	ⒶⒷⒸⒹⒺ	60	ⒻⒼⒽⒿⓀ

MATHEMATICS TEST

60 Minutes – 60 Questions

DIRECTIONS: Solve each problem, choose the correct answer, and then fill in the corresponding oval on your answer document.

Do not linger over problems that take too much time. Solve as many as you can; then return to the others in the time you have left for this test.

You are permitted to use a calculator on this test. You many use your calculator for any problems you choose, but some of the problems may best be done without using a calculator.

Note: Unless otherwise stated, all of the following should be assumed.

1. Illustrative figures are NOT necessarily drawn to scale.
2. Geometric figures lie in a plane.
3. The word line indicates a straight line.
4. The word average indicates arithmetic mean.

---

1. A perfect square is an integer that is a square of an integer. The integers h, k, and m are 16, 21 and 25, not necessarily in that order, and each integer satisfies the following condition.

- k is a perfect square.
- m is an odd integer.
- $h < k$

Which of the following statements is TRUE?

A. $h > m$
B. $m > k$
C. m is a perfect square.
D. m is a multiple of 3.
E. h is a multiple of 3.

2. The length of one piece of rope is 5 inches less than 3 times the length of a shorter piece. If the length of the longer piece is 28 inches, what is the length, in inches, of the shorter piece?

F. 9
G. 11
H. 13
J. 15
K. 17

3. If $k > 0$, which of the following is closest in value to $0.73k$?

A. $\frac{k}{7}$
B. $\frac{k}{3}$
C. $\frac{3k}{4}$
D. $\frac{k}{2}$
E. $\frac{3k}{7}$

SUGAR CONTENT OF DEFFERENT JUICES

Juice		Size	Amount of Sugar
Apple		10 oz.	60 mg
Grape		8 oz.	70 mg
Orange		12 oz.	45 mg

4. Based on the table above, which of the following lists the beverages from least to greatest sugar content per ounce?

F. A, G, O
G. A, O, G
H. G, A, O
J. G, O, A
K. O, A, G

5. If Ben collected n coins every day for 3 weeks, how many coins did he collect?

A. $n + 3$
B. $n - 3$
C. $3n$
D. $7n$
E. $21n$

6. Of 8 balls in a box, 3 are red and 5 are blue. Mike chooses 2 of these balls at random. If the first ball chosen is red, what is the probability that the second ball chosen is also red?

F. $\frac{3}{8}$
G. $\frac{3}{28}$
H. $\frac{1}{4}$
J. $\frac{5}{8}$
K. $\frac{5}{7}$

GO ON TO THE NEXT PAGE.

3, 4, 5, 6, 7, 8, 9

7. How many pairs of different numbers can be chosen from the list above so that the product of the two numbers is even? (Note: The pair 3, 4 is the same as the pair 4, 3.)

A. 6
B. 9
C. 10
D. 15
E. 21

8. If $\frac{a}{k} = b + 1$, which of the following must equal k?

F. a
G. $a + 1$
H. $b + 1$
J. $\frac{a}{b+1}$
K. $\frac{a+1}{b}$

9. Ann, Barbara and Cindy can wear a hat colored yellow, pink and blue. How many different combinations are possible?

A. 4
B. 6
C. 9
D. 12
E. 18

10. If eight bags of flour with an average (arithmetic mean) weight of 10 pounds are put onto a shelf with ten 8-pound bags of flour, what is the average weight, in pounds, of all 18 bags on the shelf?

F. 8
G. 9
H. 10
J. 11
K. 12

11. Which one of the following cannot be a factor of 48?

A. 2
B. 4
C. 6
D. 7
E. 24

12. If $n > 0$, for what value of n will $n^2 + 4 = 40$?

F. 3
G. 4
H. 6
J. 8
K. 16

13. If x is a positive even integer and $40 < 4x - 9 < 50$, what is the value of $4x - 9$?

A. 45
B. 46
C. 47
D. 48
E. 49

14. In n can be any odd integer such that $5 < \sqrt{n} < 8$, what is the difference between the largest possible value of n and the smallest possible value of n?

F. 2
G. 4
H. 36
J. 62
K. 64

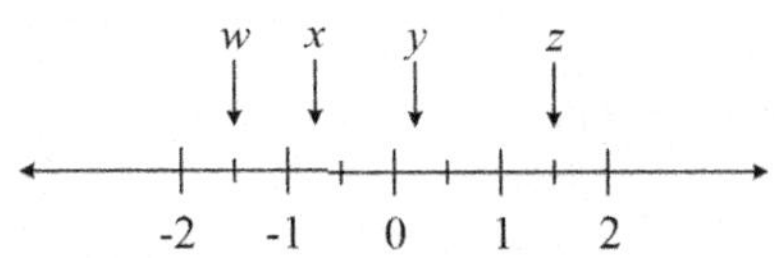

15. The letters w, x, y, and z stand for numbers on the number line above. Which of the following products is smallest?

A. wx
B. wy
C. wz
D. xy
E. yz

16. If the sum of the perimeters of two congruent squares is 40, what is the area of each square?

F. 2
G. 4
H. 5
J. 10
K. 25

17. In a survey, 100 people were asked about two television programs, A and B. Of the people surveyed, 65 watch program A, 54 watch program B, and 34 watch both programs. How many of the people surveyed watch neither of them?

A. 15
B. 20
C. 27
D. 29
E. 36

GO ON TO THE NEXT PAGE.

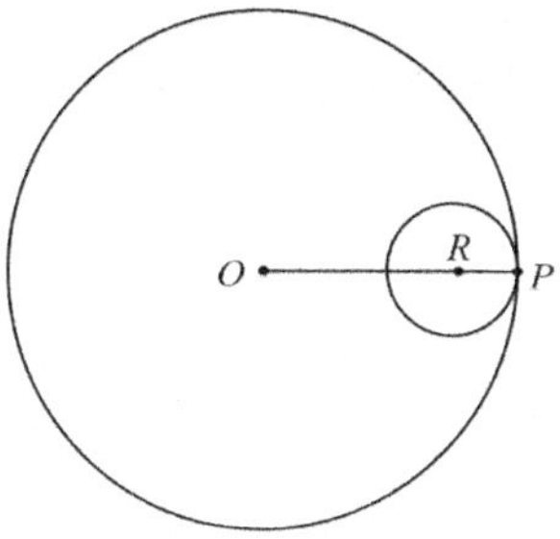

Note: Figure not drawn to scale.

18. In the figure above, point R lies on segment $\overline{OP}$. The area of the circle with center O is 144π, and the area of the circle with center R is 9π. What is the length of segment $\overline{OR}$?

F. 12
G. 9
H. 6
J. 3
K. 2

19. The volume of a rectangular box is 1 cubic foot. If the length of the box is $\frac{1}{4}$ foot and the width is $\frac{1}{2}$ foot, what is the height?

A. $\frac{1}{4}$ ft.
B. $\frac{1}{2}$ ft.
C. 2 ft.
D. 4 ft.
E. 8 ft.

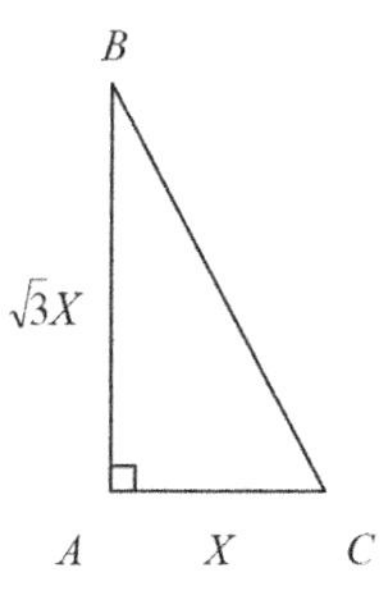

Note: Figure not drawn to scale.

20. In right triangle ABC above, what is the length of side $\overline{BC}$?

F. $2X$
G. $3X$
H. $\sqrt{5}X$
J. $\sqrt{3}X$
K. $6X$

21. If $|a|$ and $|b|$ are integers less than 10 and $\left|\frac{a}{b}\right|$ is equivalent to $\frac{3}{4}$, how many values of a are possible?

A. One
B. Two
C. Three
D. Four
E. Five

22. During a sale, a shopper pays the regular price for two bottles of milk and pays \$0.01 for the third bottle. If the regular price of the milk is \$1.87, how much per bottle does the shopper save by buying three bottles at this sale?

F. \$0.01
G. \$0.62
H. \$0.93
J. \$1.86
K. \$1.87

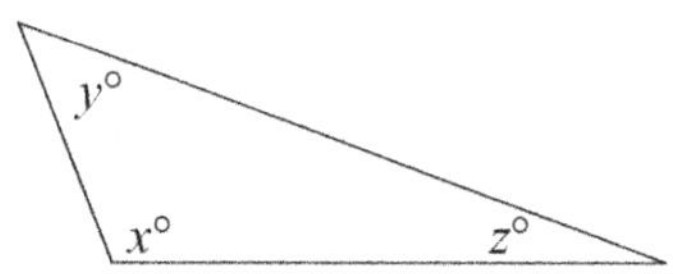

Note: Figure not drawn to scale.

23. In the figure above, $x > 90$ and $y = 2z$. If z is an integer, what is the greatest possible value of y?

A. 57
B. 58
C. 59
D. 60
E. 61

24. The sales person said, "I will give you a discount of 25 percent from the listed price. Your final price is \$299." What was the listed price?

F. \$224
G. \$349
H. \$399
J. \$449
K. \$499

25. What is the average (arithmetic mean) of 7 consecutive odd integers if the greatest of the integers is n?

A. $n-3$
B. $n-4.5$
C. $n-5$
D. $n-5.5$
E. $n-6$

GO ON TO THE NEXT PAGE.

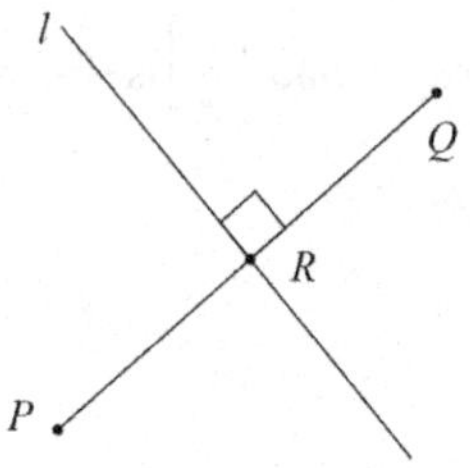

26. In the figure above, line l is a perpendicular bisector of line segment PQ and $PQ = 8$. How many circles of radius $2\sqrt{5}$ having their centers on l can be drawn that pass through both P and Q?

F. None
G. One
H. Two
J. Three
K. More than three

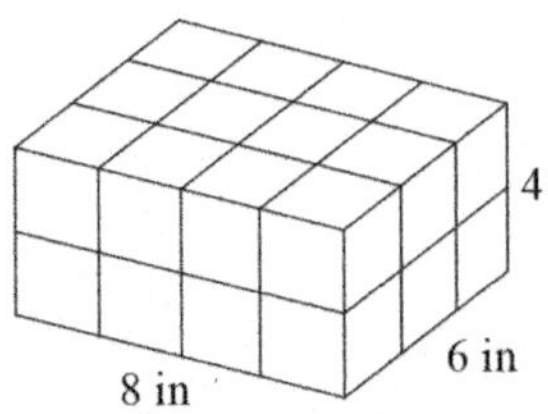

27. What is the surface area, in square inches, of the rectangular solid above?

A. 52
B. 104
C. 156
D. 208
E. 260

28. A summer camp cabin supply cabinet stocks numerals to use for a cabin room numbers. If the stock consists of 10 each of the numerals 0, 1, 2, 3, 4, 5, 6, 7, 8, and 9, how many consecutive room numbers, beginning with room number 1, can be formed?

F. 17
G. 18
H. 19
J. 22
K. 23

29. The sum of the positive even integers less than or Equal to 50 is subtracted from the sum of the positive odd integers less than 50. What is the resulting difference?

A. 20
B. 23
C. 25
D. 27
E. 28

30. In $\triangle ABC$, the measure of $\angle ABC$ is 30° and $AC < BC < AB$. If the measure of $\angle BAC$ is $x°$, what is one possible value of x?

F. 30°
G. 45°
H. 75°
J. 90°
K. 105°

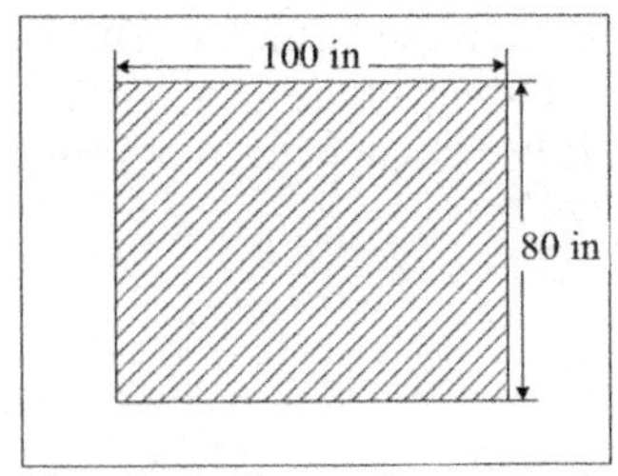

31. The shaded area shown in the figure above is going to be covered by square tiles that are 4 inches by 4 inches. How many tiles would be used to cover the shaded area?

A. 420
B. 440
C. 460
D. 480
E. 500

32. A bottle filled with juice weighs 10 pounds. When $\frac{1}{3}$ of the juice remains in the bottle, it weighs 4 pounds. What is the weight of the empty bottle, in pounds?

F. 0.75
G. 0.9
H. 1
J. 1.05
K. 1.15

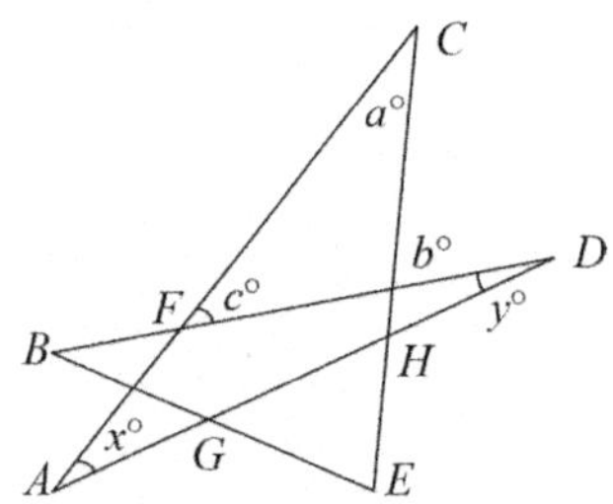

Note: Figure not drawn to scale.

33. In the figure above, AC, CE, EB, BD, and DA are line segments. If $a = 30$ and $b = 80$, what is the value of $x + y$?

A. 35°
B. 45°
C. 50°
D. 60°
E. 65°

GO ON TO THE NEXT PAGE.

34. If the product of five integers is even, at most how many of these five integers could be odd?

F. 1
G. 2
H. 3
J. 4
K. 5

35. When a number is multiplied by itself, the result is equal to 13 less than 7 times twice the number. What is one possible value of the number?

A. 5
B. 8
C. 10
D. 12
E. 13

36. A certain triangle has exactly two angles that have the same measure. If the lengths of two of the sides of the triangle are 20 and 40, what is the possible value for the perimeter of the triangle?

F. 70
G. 75
H. 80
J. 85
K. 90

[A] [B] [C] [D] [E]

37. If the 5 cards shown above are placed in a row so that [B] and [C] must be next to each other, how many different arrangements are possible?

A. 36
B. 40
C. 44
D. 48
E. 54

38. For nonzero numbers a and b, let the operations ⊠ and ⊙ be defined by $a \boxtimes b = \frac{a - b}{ab}$ and $a \odot b = \frac{ab}{a+b}$. What is the value of $5 \boxtimes (2 \odot 3)$?

F. 0.367
G. 0.433
H. 0.567
J. 0.633
K. 0.767

39. If $x^2 + 3x - 10 = 0$ and $x^2 + 7x + 10 = 0$, what is the value of x?

A. -5
B. 2
C. -2
D. 5
E. It cannot be determined from the information given.

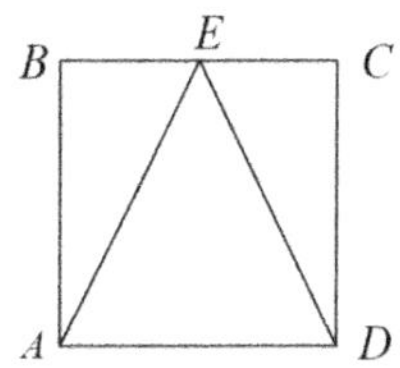

Note: Figure not drawn to scale.

40. In the figure above, the area of square $ABCD$ is 4. What is the perimeter of isosceles triangle AED?

F. 2
G. $2+\sqrt{3}$
H. $2+2\sqrt{3}$
J. $2+\sqrt{5}$
K. $2+2\sqrt{5}$

41. If the average (arithmetic mean) of x and y is 10 and the average of x, y, and z is 20, what is the value of z?

A. 40
B. 50
C. 60
D. 70
E. 80

42. If the lengths of the sides of a right triangle are consecutive even integers, which of the following can be the perimeter of the triangle?

F. 12
G. 18
H. 24
J. 30
K. 36

43. The probability of choosing a black ball from a box containing only black balls and white balls is $\frac{4}{7}$. If there are 12 white balls in the box, how many black balls are in the box?

A. 6
B. 8
C. 9
D. 12
E. 16

GO ON TO THE NEXT PAGE.

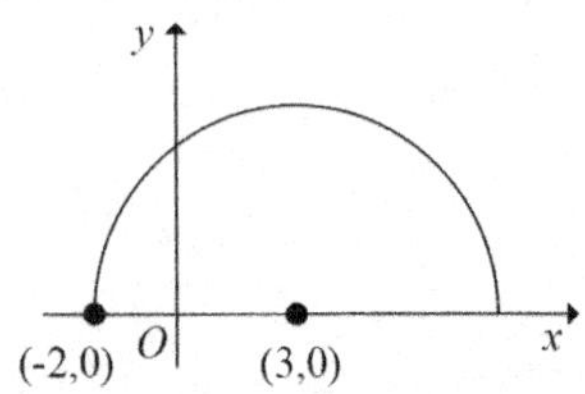

44. In the semicircle above, the center is at (3,0). Which of the following are x-coordinates of two points on this semicircle whose y-coordinates are equal?

F. -1 and 6
G. -1 and 8
H. 0 and 6
J. 2 and 6
K. 2 and 8

$32{,}032 = 14 \times 52 \times 44$

45. Based on the information above, what is the greatest prime factor of 32,032?

A. 3
B. 5
C. 7
D. 11
E. 13

46. If n is an odd integer, x is the next odd integer greater than n, and y is the next even integer greater than x, what is the value of xy in terms of n?

F. $2n + 4$
G. $2n + 5$
H. $n^2 + 3n + 2$
J. $n^2 + 4n + 3$
K. $n^2 + 5n + 6$

47. Raymond, working at a toy store, is paid k dollars each day. From this amount he spends $\frac{k}{4}$ dollars per day. In terms of k, how many days will it take Raymond to save \$1,000?

A. $\frac{k}{4{,}000}$
B. $\frac{k}{1{,}000}$
C. $\frac{1{,}000}{k}$
D. $\frac{4{,}000}{3k}$
E. $\frac{4{,}000}{k}$

48. Three points not on a line are the vertices of a certain triangle. If the lengths of the sides of the triangle are consecutive odd integers, which of the following CANNOT be the length of the longest side of the triangle?

F. 5
G. 7
H. 9
J. 11
K. 13

49. If $x = 2t^2$ and $t^3 = x$, where t is a positive integer, what is the value of x?

A. 2
B. 8
C. 6
D. 5
E. 4

50. The first term in the sequence of numbers shown above is 7. Each even-numbered term is -1 times the previous term and each odd-numbered term, after the first, is 3 less than the previous term. What is the 47th term of the sequence?

F. -10
G. -7
H. -1
J. 7
K. 10

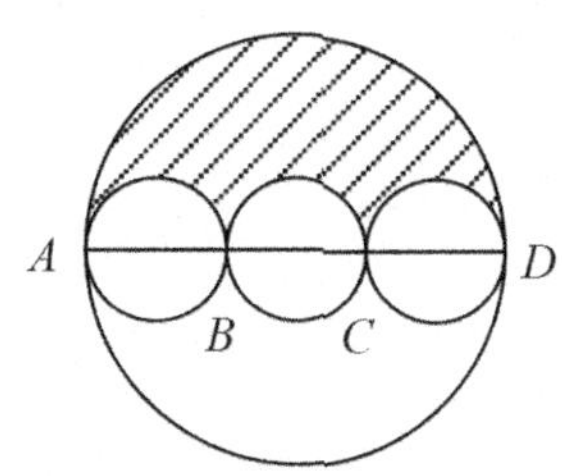

Note: Figure not drawn to scale.

51. AB, BC, CD and AD are diameters of the four circles shown above. If $AB = BC = CD = 2$, what is the area of the shaded region?

A. 12π
B. 6π
C. $\frac{9}{2}\pi$
D. 3π
E. 2π

GO ON TO THE NEXT PAGE.

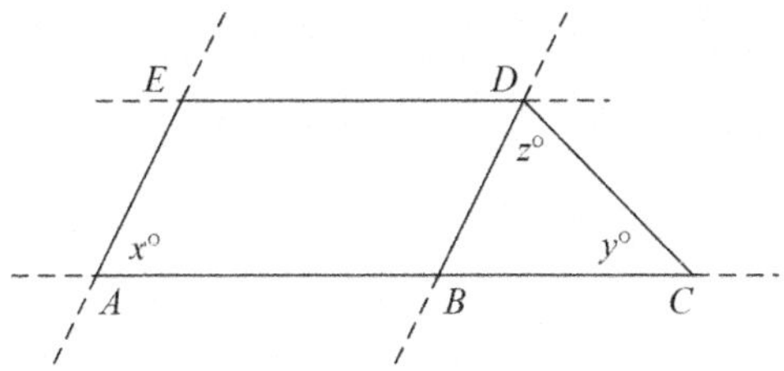

Note: Figure not drawn to scale.

52. In the figure above, if $x + y + z < 180$ and AC is parallel to ED, which of the following must be true?

F. Line AE and BD will intersect when extended.
G. Line BD is not perpendicular to line DC.
H. $x = z$
J. $y > x$
K. $y < x$

7, -7, -10, ...

53. When the radius of a given circle is lengthened by 1 inch, the area is increased by 9π square inches. What is the length, in inches, of the radius of the original circle?

A. 4
B. 6
C. 8
D. 9
E. 10

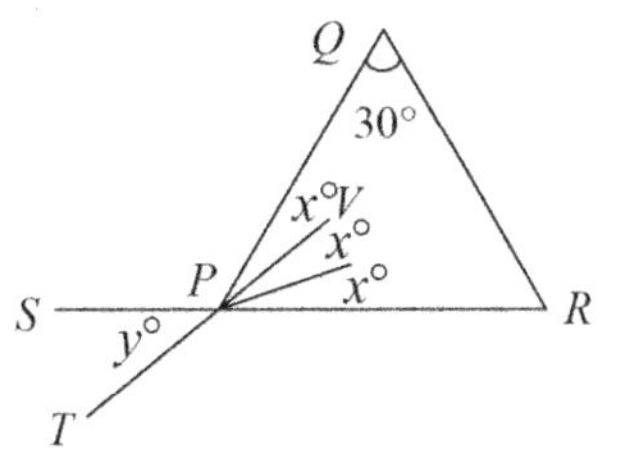

54. In the figure above, ΔPQR is an isosceles triangle with the vertex angle $\theta = 30°$, and SR and TV are line segments. What is the value of $x + y$?

F. 60
G. 75
H. 20
J. 45
K. 15

55. If A, B, C, D and F are 5 distinct points on the circumference of a circle, how many different chords can be drawn using any 2 of the 5 points?

A. 6
B. 8
C. 10
D. 12
E. 24

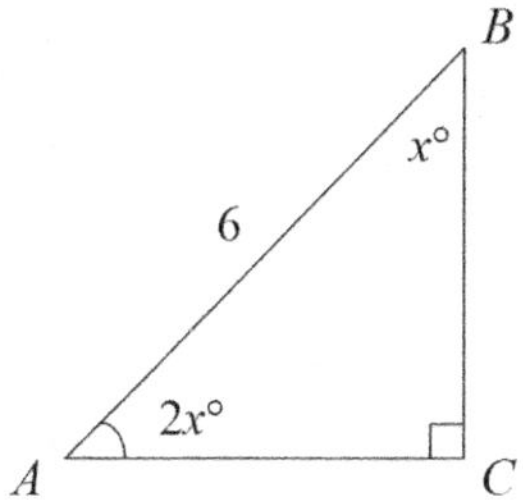

56. In Figure shown above, what is the perimeter?

F. $9 + 3\sqrt{3}$
G. 10
H. $5 + 2\sqrt{3}$
J. $10 + \sqrt{3}$
K. $12 + \sqrt{2}$

57. If $x = \sqrt[3]{t}$ and $y = t^2$, then $\frac{y}{x^5} =$

A. $t^{-\frac{1}{3}}$
B. $t^{\frac{1}{3}}$
C. $t^{\frac{5}{6}}$
D. $t^{\frac{6}{5}}$
E. $t^{\frac{10}{3}}$

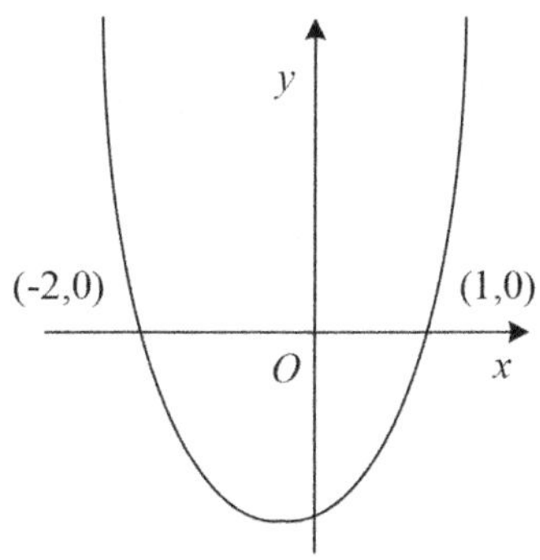

58. Which of the following could be the equation of the parabola in the figure shown above?

F. $y = (x - 2)(x - 3)$
G. $y = (x + 2)(x + 3)$
H. $y = (x + 2)(x - 3)$
J. $y = (x + 2)(x - 1)$
K. $x = (y + 2)(y - 3)$

GO ON TO THE NEXT PAGE.

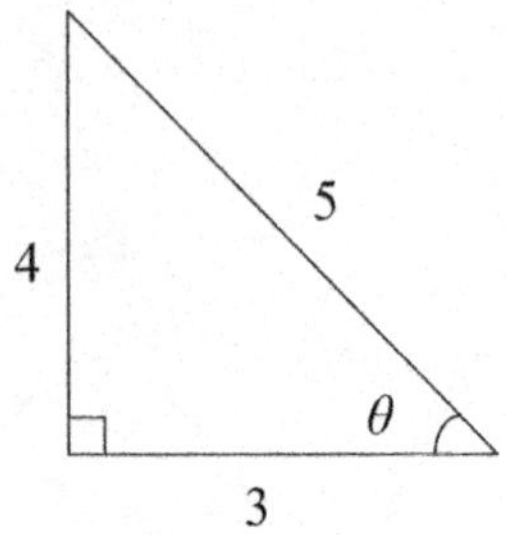

59. In above figure, $\frac{1}{\tan\theta} + \sec\theta =$

A. 0.75
B. 1.20
C. 1.43
D. 2.74
E. 2.42

60. For what value of x will $f(x) = 3x^2 - 4x + 5$ have the minimum value?

F. -2
G. $-\frac{1}{3}$
H. 0
J. $\frac{2}{3}$
K. 1

END OF TEST 5
STOP! DO NOT TURN THE PAGE UNTIL TOLD TO DO SO.
DO NOT RETURN TO THE PREVIOUS TEST.

Answer Key
Model Test No. 05

1	D	16	K	31	E	46	K
2	G	17	A	32	H	47	E
3	C	18	G	33	C	48	F
4	K	19	E	34	J	49	B
5	E	20	F	35	E	50	F
6	G	21	D	36	H	51	D
7	D	22	G	37	D	52	F
8	J	23	B	38	J	53	A
9	C	24	H	39	A	54	G
10	G	25	E	40	K	55	C
11	D	26	H	41	A	56	F
12	H	27	D	42	H	57	B
13	C	28	F	43	E	58	J
14	H	29	C	44	H	59	E
15	C	30	B	45	E	60	J

How to Score the ACT Mathematics Test

When you take an actual ACT Mathematics Test, your answer sheet will be "read" by a scanning machine that will record your responses to each question. Then a computer will compare your answers with the correct answers and produce your raw score. You get one point for each correct answer. For each wrong answer, there is no penalty. Therefore, you must guess and answer all the questions to get a higher score. This raw score is converted to a scaled score that is reported to you and to the colleges you specify.

Finding Your Scale Test Score

STEP 1: Count the number of right answers.
Enter the total here: ______________________

STEP 2: The number you obtained in Step 1 is your raw score.
Look up the conversion table for ACT-Math, and find your score.

Scaled Score Conversion Table ACT Mathematics Test	
Scale Score	Mathematics
36	60
35	58-59
34	56-57
33	55
32	54
31	52-53
30	50-51
29	48-49
28	46-47
27	43-45
26	41-42
25	39-40
24	37-38
23	35-36
22	33-34
21	31-32
20	30
19	27-29
18	25-26
17	22-24
16	18-21
15	15-17
14	12-14
13	09-11
12	08
11	06-07
10	05
09	04
08	-
07	03
06	02
05	-
04	-
03	01
02	-
01	00

1. A perfect square is an integer that is a square of an integer. The integers h, k, and m are 16, 21 and 25, not necessarily in that order, and each integer satisfies the following condition.
   - k is a perfect square.
   - m is an odd integer.
   - $h < k$

   Which of the following statements is TRUE?

   A. $h > m$
   B. $m > k$
   C. m is a perfect square.
   D. m is a multiple of 3.
   E. h is a multiple of 3.

Since k is a perfect square, it is either 16 or 25.
But $h < k$, thus h = 16, $k = 25$.
Also, m is an odd integer, thus $m = 21$.
So, $h = 16$, $m = 21$, $k = 25$.

Ans. D

2. The length of one piece of rope is 5 inches less than 3 times the length of a shorter piece. If the length of the longer piece is 28 inches, what is the length, in inches, of the shorter piece?

   F. 9 **G.** 11 **H.** 13 **J.** 15 **K.** 17

$L = 3S - 5$, where L is the length of longer piece, and S is the length of a shorter one. Now, $L = 28$.
$\therefore 28 = 3S - 5$, or $3S = 33$.
$S = 11$

Ans. G

3. If $k > 0$, which of the following is closest in value to $0.73k$?

   A. $\frac{k}{7}$ **B.** $\frac{k}{3}$ **C.** $\frac{3k}{4}$ **D.** $\frac{k}{2}$ **E.** $\frac{3k}{7}$

$\frac{3}{4}k = 0.75k \approx 0.73k$

Ans. C

SUGAR CONTENT OF DEFFERENT JUICES

Juice	Size	Amount of Sugar
Apple	10 oz.	60 mg
Grape	8 oz.	70 mg
Orange	12 oz.	45 mg

4. Based on the table above, which of the following lists the beverages from least to greatest sugar content per ounce?

   F. A, G, O **G.** A, O, G **H.** G, A, O
   J. G, O, A **K.** O, A, G

The sugar content of each juice is as follows:

Apple $= \frac{60 \text{ mg}}{10 \text{ oz.}} = 6$ mg/oz.

Grape $= \frac{70 \text{ mg}}{8 \text{ oz.}} = 8.75$ mg/oz.

Orange $= \frac{45 \text{ mg}}{12 \text{ oz.}} = 3.75$ mg/oz.

$\therefore 0 < A < G$

Ans. K

5. If Ben collected n coins every day for 3 weeks, how many coins did he collect?

   A. $n + 3$
   B. $n - 3$
   C. $3n$
   D. $7n$
   E. $21n$

$n \times 7$ days $\times$ 3 weeks $= 21n$

Ans. E

6. Of 8 balls in a box, 3 are red and 5 are blue. Mike chooses 2 of these balls at random. If the first ball chosen is red, what is the probability that the second ball chosen is also red?

F. $\frac{3}{8}$ **G.** $\frac{3}{28}$ **H.** $\frac{1}{4}$ **J.** $\frac{5}{8}$ **K.** $\frac{5}{7}$

R = 3
B = 5
T = 8

$P(R - R) = \frac{3}{8} \times \frac{2}{7} = \frac{6}{56} = \frac{3}{28}$

Ans. G

3, 4, 5, 6, 7, 8, 9

7. How many pairs of different numbers can be chosen from the list above so that the product of the two numbers is even? (Note: The pair 3, 4 is the same as the pair 4, 3.)

A. 6 **B.** 9 **C.** 10 **D.** 15 **E.** 21

$3 \times 4, 3 \times 6, 3 \times 8,$
$4 \times 5, 4 \times 6, 4 \times 7, 4 \times 8, 4 \times 9,$
$5 \times 6, 5 \times 8,$
$6 \times 7, 6 \times 8, 6 \times 9,$
$7 \times 8,$
8×9
Total = 15

Ans. D

8. If $\frac{a}{k} = b + 1$, which of the following must equal k?

F. a **G.** $a + 1$ **H.** $b + 1$ **J.** $\frac{a}{b+1}$ **K.** $\frac{a+1}{b}$

$\frac{a}{k} = b + 1$, or $\frac{k}{a} = \frac{1}{b+1}$ $\therefore k = \frac{a}{b+1}$

Ans. J

9. Ann, Barbara and Cindy can wear a hat colored yellow, pink and blue. How many different combinations are possible?

A. 4 **B.** 6 **C.** 9 **D.** 12 **E.** 18

3 people × 3 hats = 9

Ans. C

10. If eight bags of flour with an average (arithmetic mean) weight of 10 pounds are put onto a shelf with ten 8-pound bags of flour, what is the average weight, in pounds, of all 18 bags on the shelf?

F. 8 **G.** 9 **H.** 10 **J.** 11 **K.** 12

$\text{Avg.} = \frac{(8 \times 10) + (10 \times 8)}{18 \text{ bags}} = \frac{160}{18} \approx 9$

Ans. G

11. Which one of the following cannot be a factor of 48?

A. 2 **B.** 4 **C.** 6 **D.** 7 **E.** 24

$48 = 2 \times 2 \times 2 \times 2 \times 3$
$\therefore$ 7 cannot be a factor.

Ans. D

12. If $n > 0$, for what value of n will $n^2 + 4 = 40$?

F. 3 **G.** 4 **H.** 6 **J.** 8 **K.** 16

$n^2 = 36 \quad \therefore n = 6$

Ans. H

13. If x is a positive even integer and $40 < 4x - 9 < 50$, what is the value of $4x - 9$?

A. 45 **B.** 46 **C.** 47 **D.** 48 **E.** 49

$40 < 4x - 9 < 50$
$\therefore 49 < 4x < 59$, or $\frac{49}{4} < x < \frac{59}{4} \rightarrow 12.25 < x < 14.75$
Since x is even integers, we get $x = 14$.
$\therefore 4x - 9 = 4(14) - 9 = 56 - 9 = 47$

Ans. C

14. In n can be any odd integer such that $5 < \sqrt{n} < 8$, what is the difference between the largest possible value of n and the smallest possible value of n?

F. 2 **G.** 4 **H.** 36 **J.** 62 **K.** 64

$5 < \sqrt{n} < 8$
$\therefore 25 < n < 64$
But n = odd integer
$\therefore$ the smallest $n = 27$, and the largest $n = 63$
$\therefore 63 - 27 = 36$

Ans. H

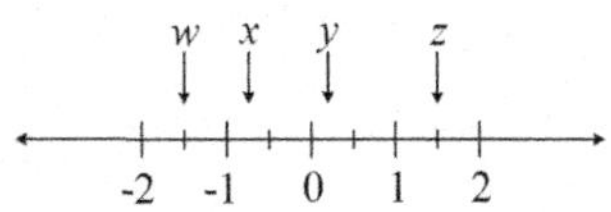

15. The letters w, x, y, and z stand for numbers on the number line above. Which of the following products is smallest?

A. wx **B.** wy **C.** wz **D.** xy **E.** yz

To get the product of two numbers to be the smallest, the sign of the numbers must be opposite to get a negative number, while the absolute value of the product must be the largest!! Therefore, wz = the smallest.

Ans. C

16. If the sum of the perimeters of two congruent squares is 40, what is the area of each square?

F. 2 **G.** 4 **H.** 5 **J.** 10 **K.** 25

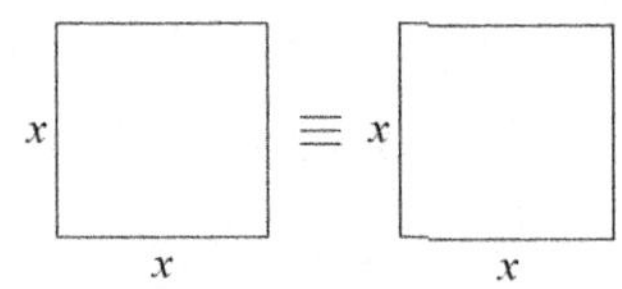

$P_1 = 4x$ $P_2 = 4x$
$\therefore P_1 + P_2 = 8x = 40$
$\therefore x = 5$
$\therefore$ Area $= 5^2 = 25$

Ans. K

17. In a survey, 100 people were asked about two television programs, A and B. Of the people surveyed, 65 watch program A, 54 watch program B, and 34 watch both programs. How many of the people surveyed watch neither of them?

A. 15 **B.** 20 **C.** 27 **D.** 29 **E.** 36

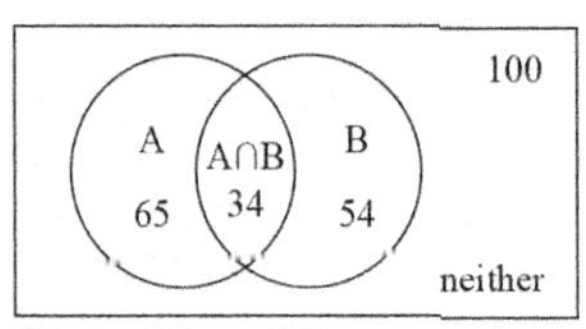

Given Venn Diagram above, we get neither N = 100 – (AUB) = 100 – [A + B – (A∩B)] = 100 – (65 + 54 – 34) = 15

Ans. A

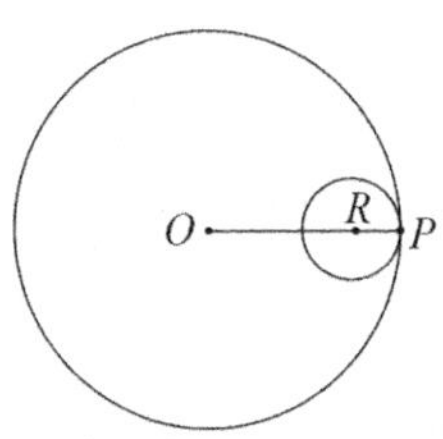

18. In the figure above, point R lies on segment $\overline{OP}$. The area of the circle with center O is 144π, and the area of the circle with center R is 9π. What is the length of segment $\overline{OR}$?

F. 12 **G.** 9 **H.** 6 **J.** 3 **K.** 2

Let the radius of the bigger circle be, $\overline{OP} = x$, and the radius of the smaller circle, $\overline{RP} = y$.
The, $144\pi = \pi x^2$
$\therefore x = 12$, and $9\pi = \pi y^2$
$\therefore y = 3$
$\therefore 12 - 3 = 9$

Ans. G

19. The volume of a rectangular box is 1 cubic foot. If the length of the box is $\frac{1}{4}$ foot and the width is $\frac{1}{2}$ foot, what is the height?

A. $\frac{1}{4}$ ft. **B.** $\frac{1}{2}$ ft. **C.** 2 ft. **D.** 4 ft. **E.** 8 ft.

$V = 1 = \frac{1}{2} \times \frac{1}{4} \times h$

$\therefore h = 8$

Ans. E

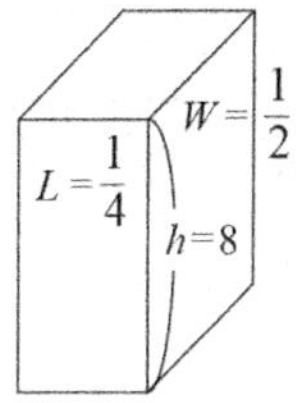

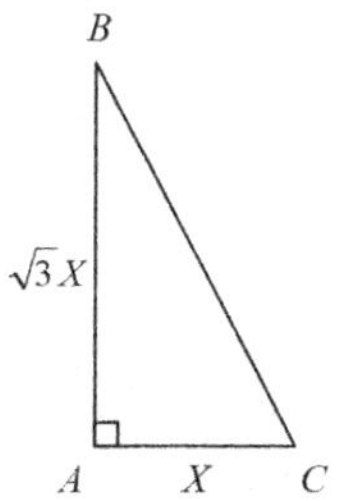

20. In right triangle ABC above, what is the length of side $\overline{BC}$?

F. $2X$ **G.** $3X$ **H.** $\sqrt{5}X$ **J.** $\sqrt{3}X$ **K.** $6X$

$\overline{BC} = \sqrt{(x^2 + (\sqrt{3}x)^2} = \sqrt{4x^2} = 2x$

Ans. F

21. If $|a|$ and $|b|$ are integers less than 10 and $\left|\frac{a}{b}\right|$ is equivalent to $\frac{3}{4}$, how many values of a are possible?

A. One **B.** Two **C.** Three **D.** Four **E.** Five

The only possible values for $\left|\frac{a}{b}\right| = \frac{3}{4}$ or $\frac{6}{8}$, when $|a| = 3$ or 6.

$\therefore a = \pm 3, \pm 6$

Ans. D

22. During a sale, a shopper pays the regular price for two bottles of milk and pays \$0.01 for the third bottle. If the regular price of the milk is \$1.87, how much per bottle does the shopper save by buying three bottles at this sale?

F. \$0.01 **G.** \$0.62 **H.** \$0.93 **J.** \$1.86 **K.** \$1.87

The regular price for 3 bottles is, $3 \times \$1.87 = \$5.61$

Also, the sale price is $(2 \times \$1.87) + (\$0.01) = \$3.75$

$\therefore$ The saving for 3 bottles is $(\$5.61 - \$3.75) = \$1.86$

$\therefore$ each bottle saving $= \$1.86 \div 3 = \$0.62$

Ans. G

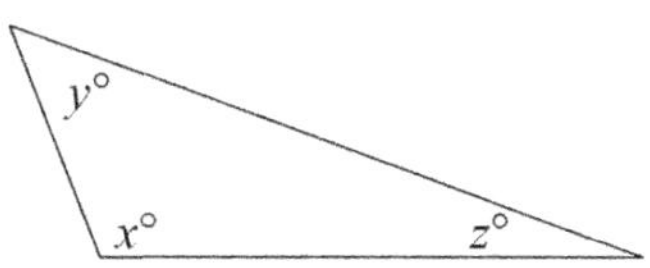

23. In the figure above, $x > 90$ and $y = 2z$. If z is an integer, what is the greatest possible value of y?

A. 57 **B.** 58 **C.** 59 **D.** 60 **E.** 61

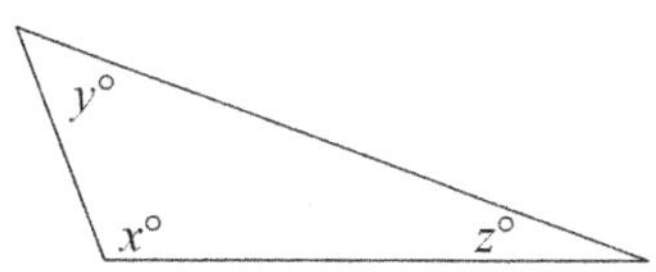

To get the max of angle y, we must have x to be the min,

$\therefore$ Given $x > 90$, $x = 91$

But z is an integer, and $y + z = 2z + z = 3z$ must also be satisfied with $(180 - x)$.

$\therefore x \neq 91$, since z cannot be integer.

The possible minimum value for x should be 93, so that $3z = 87$.

$\therefore z = 29$ $\therefore y = 58$

Ans. B

24. The sales person said, "I will give you a discount of 25 percent from the listed price. Your final price is \$299." What was the listed price?

F. \$224 **G.** \$349 **H.** \$399 **J.** \$449 **K.** \$499

Let L = listed price, then $\$299 = (1 - 0.25)L$

$\therefore L = \$\frac{299}{0.75} = \$399$

Ans. H

25. What is the average (arithmetic mean) of 7 consecutive <u>odd</u> integers if the greatest of the integers is n?

A. $n-3$ **B.** $n-4.5$ **C.** $n-5$
D. $n-5.5$ **E.** $n-6$

Let's just make $n = 15$, then 15, 13, 11, 9, 7, 5, 3 are the 7 consecutive odd integers. $\therefore$ Avg. $= \dfrac{\text{sum of all}}{\text{7 numbers}} = \dfrac{63}{7} = 9$
Now, try answer choice **A**, **B** … with $n = 15$ to get Avg. = 9.
The only choice is **E**, $n - 6 = 15 - 6 = 9$.

Ans. E

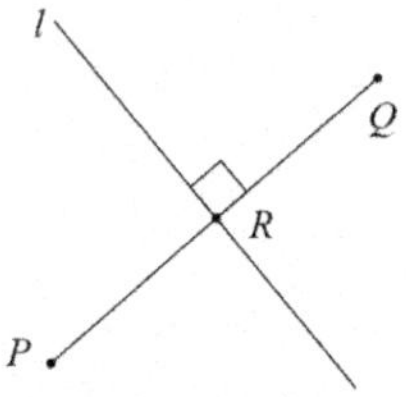

26. In the figure above, line l is a perpendicular bisector of line segment PQ and $PQ = 8$. How many circles of radius $2\sqrt{5}$ having their centers on l can be drawn that pass through both P and Q?

A. None
B. One
C. Two
J. Three
K. More than three

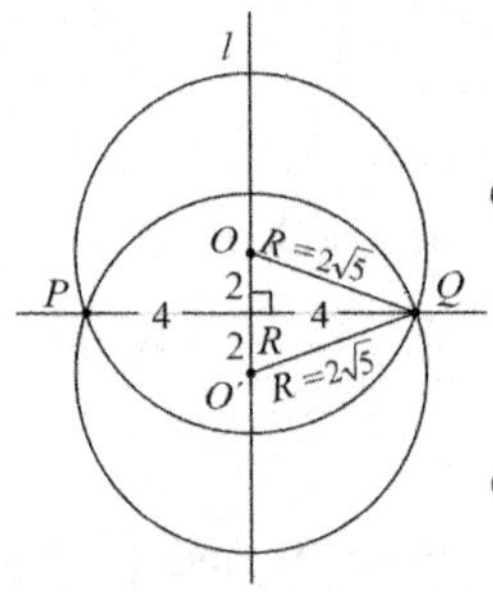

circle with center O and $r = 2\sqrt{5}$

circle with center O' and $r = 2\sqrt{5}$

Given information that $\overline{PQ} = 8$ and l is a P.B. line, we get $\overline{PR} = \overline{RQ} = 4$. Now, to get $r = 2\sqrt{5} = \sqrt{20}$, we have

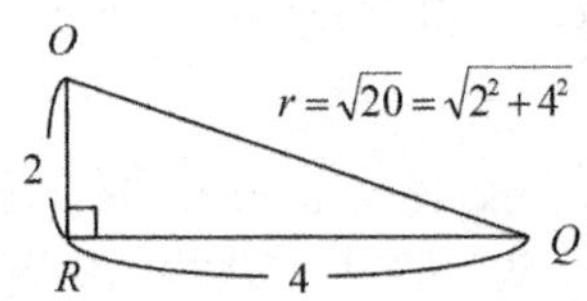

Therefore, we get 2 circles with center on line l with point $O = (0, 2)$ and $O' = (0, -2)$.

Ans. H

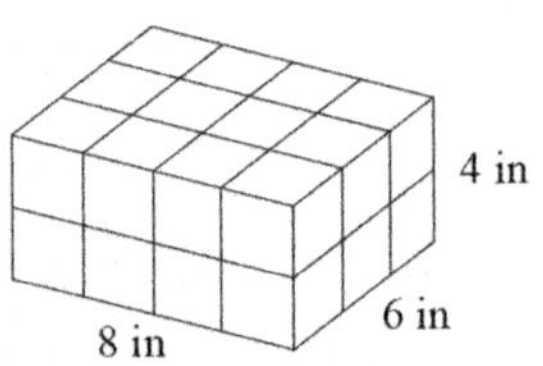

27. What is the surface area, in square inches, of the rectangular solid above?

A. 52 **B.** 104 **C.** 156 **D.** 208 **E.** 260

S.A. $= 2[(8 \times 6) + (4 \times 6) + (4 \times 8)] = 208$

Ans. D

28. A summer camp cabin supply cabinet stocks numerals to use for a cabin room numbers. If the stock consists of 10 each of the numerals 0, 1, 2, 3, 4, 5, 6, 7, 8, and 9, how many consecutive room numbers, beginning with room number 1, can be formed?

F. 17 **G.** 18 **H.** 19 **J.** 22 **K.** 23

The cabin room numbers will start from 1 and it goes on until it runs out of the numeral 1. That is,

$\boxed{1}\,\boxed{2}\,\boxed{3} \dots \boxed{9}$

$\boxed{1}\boxed{0}, \boxed{1}\boxed{1}, \boxed{1}\boxed{2} \dots \boxed{1}\boxed{7}.$

As we may count $\boxed{1}$'s above, $\boxed{1}$ runs out on $\boxed{1}\boxed{7}$.

Ans. F

29. The sum of the positive even integers less than or Equal to 50 is subtracted from the sum of the positive odd integers less than 50. What is the resulting difference?

A. 20 **B.** 23 **C.** 25 **D.** 27 **E.** 28

$(2 + 4 + 6 + \dots + 48 + 50)$
$\underline{-(1 + 3 + 5 + \dots + 47 + 49)}$
$(1 + 1 + 1 + \dots + 1 + 1)$, which is 25 of 1's.

Ans. C

30. In $\triangle ABC$, the measure of $\angle ABC$ is 30° and $AC < BC < AB$. If the measure of $\angle BAC$ is $x°$, what is one possible value of x?

F. 30°
G. 45°
H. 75°
J. 90°
K. 105°

Since we have $AC < BC < AB$, the angles become $\angle B < \angle A < \angle C$. Therefore, $\angle A = x°$ becomes the middle one.
Now, $\angle A + \angle C = 180° - 30° = 150°$
∴ The angle x must be greater than the smallest angle 30°, but less than $\frac{1}{2} \times 150°$, which is 75°.
∴ Any angle between 30° and 75°.

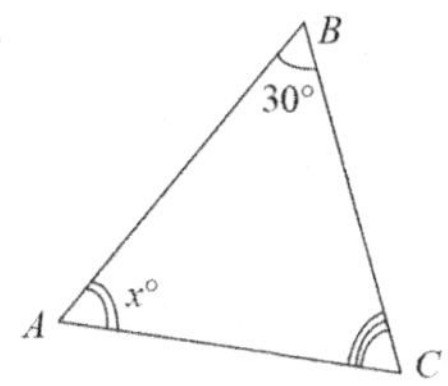

Ans. B

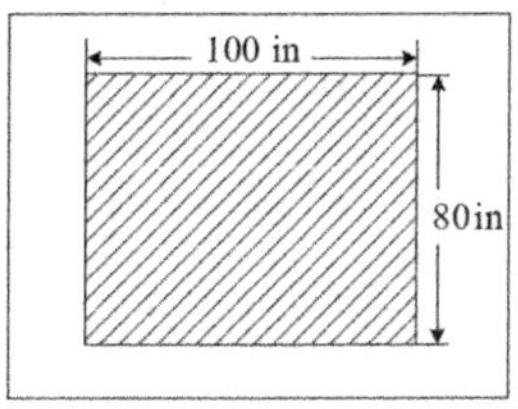

31. The shaded area shown in the figure above is going to be covered by square tiles that are 4 inches by 4 inches. How many tiles would be used to cover the shaded area?

A. 420 **B.** 440 **C.** 460 **D.** 480 **E.** 500

The total shaded are = $100 \times 80 = 8000$.
∴ $8000 \div (4 \times 4) = 500$

Ans. E

32. A bottle filled with juice weighs 10 pounds. When $\frac{1}{3}$ of the juice remains in the bottle, it weighs 4 pounds. What is the weight of the empty bottle, in pounds?

F. 0.75 **G.** 0.9 **H.** 1 **J.** 1.05 **K.** 1.15

Let B = the weight of the bottle, and J = the weight of the juice.
Then, B + J = 10 - eq ①, B + $\frac{1}{3}$J = 4 - eq ②
∴ 3 × eq ② – eq ① = 2B = 2 ∴ B = 1

Ans. H

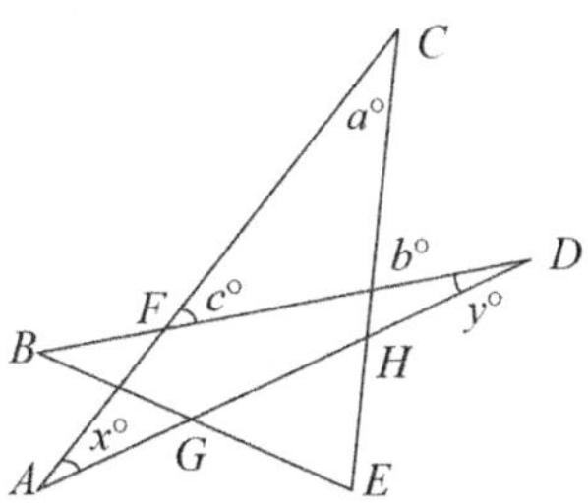

33. In the figure above, AC, CE, EB, BD, and DA are line segments. If $a = 30$ and $b = 80$, what is the value of $x + y$?

A. 35° **B.** 45° **C.** 50° **D.** 60° **E.** 65°

$\angle b = \angle a + \angle c$, because $\angle b$ is the exterior angle, which is the sum of $\angle a$ and $\angle c$.
∴ $80 = 30 + \angle c$
∴ $\angle c = 50°$
But angle $\angle c$ is also the exterior angle, which is the sum of $\angle x$ and $\angle y$.
∴ $\angle c = 50° = x + y$

Ans. C

34. If the product of five integers is even, at most how many of these five integers could be odd?

F. 1 **G.** 2 **H.** 3 **J.** 4 **K.** 5

Consider odd × odd = odd and even × odd = even.
∴ We get odd × odd × odd × odd × even = even.
∴ 4 odds

Ans. J

35. When a number is multiplied by itself, the result is equal to 13 less than 7 times twice the number. What is one possible value of the number?

A. 5 **B.** 8 **C.** 10 **D.** 12 **E.** 13

$x^2 = 7(2x) - 13$
$\therefore x^2 - 14x + 13 = 0$, $(x - 13)(x - 1) = 0$
$\therefore x = 1$ or 13

Ans. E

36. A certain triangle has exactly two angles that have the same measure. If the lengths of two of the sides of the triangle are 20 and 40, what is the possible value for the perimeter of the triangle?

F. 70 **G.** 75 **H.** 80 **J.** 85 **K.** 90

Since the triangle has exactly 2 angles with the same measure, it is an isosceles triangle as shown here.

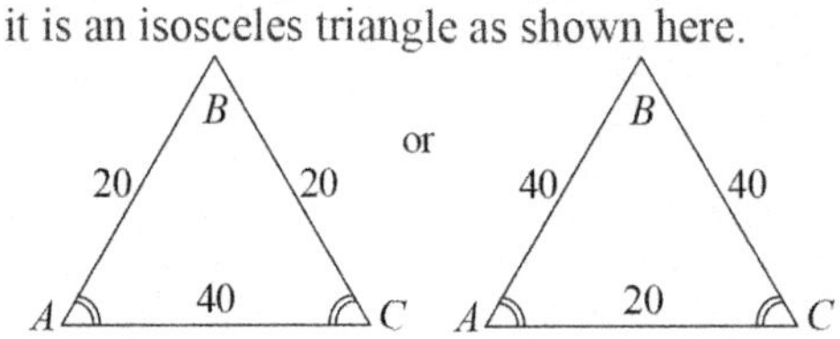

Therefore, P = 80 or 100.

Ans. H

$\boxed{A}$ $\boxed{B}$ $\boxed{C}$ $\boxed{D}$ $\boxed{E}$

37. If the 5 cards shown above are placed in a row so that $\boxed{B}$ and $\boxed{C}$ must be next to each other, how many different arrangements are possible?

A. 36 **B.** 40 **C.** 44 **D.** 48 **E.** 54

When we consider the two cards $\boxed{B}$ and $\boxed{C}$ as one group, always being together, then we have
$(\boxed{B}-\boxed{C})-\boxed{A}-\boxed{D}-\boxed{E}$,
which we have 4! ways of arrangement.
Now, there are 2! ways of placing the two cards, $(\boxed{B}-\boxed{C})$.
∴ we get 4! × 2! = 48

Ans. D

38. For nonzero numbers a and b, let the operations $\boxtimes$ and $\odot$ be defined by $a \boxtimes b = \dfrac{a-b}{ab}$ and $a \odot b = \dfrac{ab}{a+b}$.
What is the value of $5 \boxtimes (2 \odot 3)$?

F. 0.367
G. 0.433
H. 0.567
J. 0.633
K. 0.767

First, $(2 \odot 3) = \dfrac{2 \times 3}{2+3} = \dfrac{6}{5} = 1.2$

Now, $5 \boxtimes (2 \odot 3) = 5 \boxtimes 1.2 = \dfrac{5 - 1.2}{5 \times 1.2} = \dfrac{3.8}{6} = \dfrac{19}{30}$ or 0.633

Ans. J

39. If $x^2 + 3x - 10 = 0$ and $x^2 + 7x + 10 = 0$, what is the value of x?

A. -5
B. 2
C. -2
D. 5
E. It cannot be determined from the information given.

$x^2 + 3x - 10 = (x + 5)(x - 2) = 0$
$\therefore x = -5$ or 2
$x^2 + 7x + 10 = (x + 5)(x + 2) = 0$
$\therefore x = -5$ or -2
Therefore, $x = -5$ is the solution that satisfies both.

Ans. A

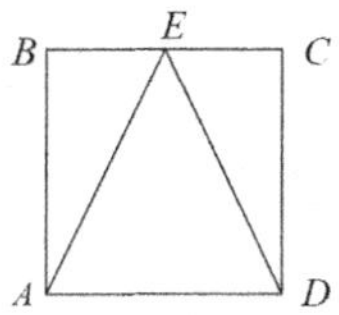

40. In the figure above, the area of square $ABCD$ is 4. What is the perimeter of isosceles triangle AED?

F. 2 **G.** $2+\sqrt{3}$ **H.** $2+2\sqrt{3}$
J. $2+\sqrt{5}$ **K.** $2+2\sqrt{5}$

Using Pythagorean Th.,
we get $\overline{AE}=\sqrt{1^2+2^2}=\sqrt{5}$
$\therefore$ Perimeter of ΔAED, $P=2+2\sqrt{5}$

Ans. K

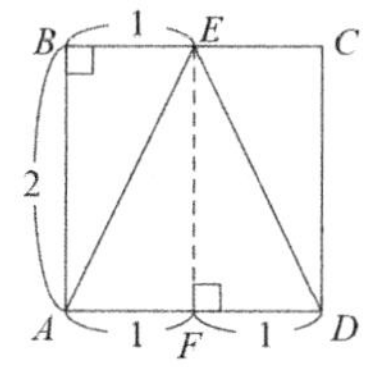

41. If the average (arithmetic mean) of x and y is 10 and the average of x, y, and z is 20, what is the value of z?

A. 40 **B.** 50 **C.** 60 **D.** 70 **E.** 80

$\frac{x+y}{2}=10 \quad \therefore x+y=20$

Also, $\frac{x+y+z}{3}=\frac{20+z}{3}=20 \quad \therefore z=40$

Ans. A

42. If the lengths of the sides of a right triangle are consecutive even integers, which of the following can be the perimeter of the triangle?

F. 12 **G.** 18 **H.** 24 **J.** 30 **K.** 36

$x^2+(x+2)^2=(x+4)^2$
$\therefore x^2+x^2+4x+4=x^2+8x+16$
$\therefore x^2-4x-12=0$ or $(x-6)(x+2)=0$
$\therefore x=-2$ or 6
Now, let $x=6$, then 6, 8 and 10.
$\therefore p=6+8+10=24$

Ans. H

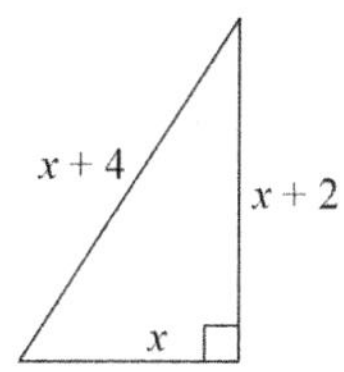

43. The probability of choosing a black ball from a box containing only black balls and white balls is $\frac{4}{7}$. If there are 12 white balls in the box, how many black balls are in the box?

A. 6 **B.** 8 **C.** 9 **D.** 12 **E.** 16

$P\text{(Black)}=\frac{x}{x+12}=\frac{4}{7}$
$\therefore 7x=4x+48$
$\therefore 3x=48$
$\therefore x=16$

Ans. E

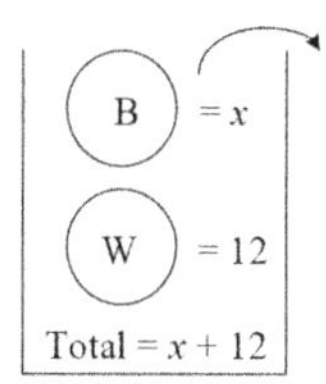

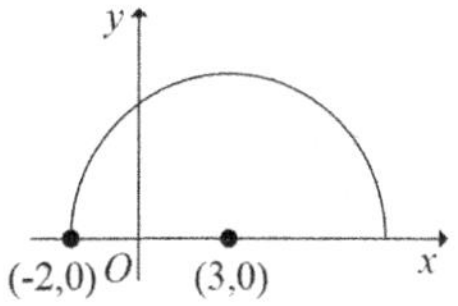

44. In the semicircle above, the center is at (3,0). Which of the following are x-coordinates of two points on this semicircle whose y-coordinates are equal?

F. -1 and 6 **G.** -1 and 8 **H.** 0 and 6
J. 2 and 6 **K.** 2 and 8

Since the semi-circle has center at (3, 0), and its radius $r=5$, we must look for the x-coordinates that are symmetric to $x=3$ line. $x=0$ and $x=6$ are the points that are symmetric to $x=3$ line.

Ans. H

$32{,}032=14\times 52\times 44$

45. Based on the information above, what is the greatest prime factor of 32,032?

A. 3 **B.** 5 **C.** 7 **D.** 11 **E.** 13

$32{,}032=14\times 52\times 44=(7\times 2)\times(2\times 2\times 13)\times(2\times 2\times 11)$
$\therefore$ The greatest prime factor is 13.

Ans. E

46. If n is an odd integer, x is the next odd integer greater than n, and y is the next even integer greater than x, what is the value of xy in terms of n?

F. $2n + 4$ **G.** $2n + 5$ **H.** $n^2 + 3n + 2$
J. $n^2 + 4n + 3$ **K.** $n^2 + 5n + 6$

$x = n + 2, y = n + 3$
$\therefore xy = (n + 2)(n + 3) = n^2 + 5n + 6$

Ans. K

47. Raymond, working at a toy store, is paid k dollars each day. From this amount he spends $\frac{k}{4}$ dollars per day. In terms of k, how many days will it take Raymond to save \$1,000?

A. $\frac{k}{4,000}$ **B.** $\frac{k}{1,000}$ **C.** $\frac{1,000}{k}$ **D.** $\frac{4,000}{3k}$ **E.** $\frac{4,000}{k}$

Raymond's saving per day is, $S = (k - \frac{k}{4}) = \frac{3k}{4}$ per day.

$\therefore \$\frac{1,000}{\frac{3k}{4}} = \frac{4,000}{3k}$ days to save \$1,000.

Ans. E

48. Three points not on a line are the vertices of a certain triangle. If the lengths of the sides of the triangle are consecutive odd integers, which of the following CANNOT be the length of the longest side of the triangle?

F. 5 **G.** 7 **H.** 9 **J.** 11 **K.** 13

Remember, for any triangle, the sum of two sides is always greater than the other!! Therefore, if we set the side length to be 1, 3 and 5, then, $1 + 3 < 5$, which cannot form a triangle.

Ans. F

49. If $x = 2t^2$ and $t^3 = x$, where t is a positive integer, what is the value of x?

A. 2 **B.** 8 **C.** 6 **D.** 5 **E.** 4

$x = 2t^2 = t^3$
$\therefore t^3 - 2t^2 = 0$
$\therefore t^2(t - 2) = 0$
$\therefore t = 0, 2$
$\therefore x = 2(2)^2 = 8$

Ans. B

7, -7, -10, ...

50. The first term in the sequence of numbers shown above is 7. Each even-numbered term is -1 times the previous term and each odd-numbered term, after the first, is 3 less than the previous term. What is the 47th term of the sequence?

F. -10 **G.** -7 **H.** -1 **J.** 7 **K.** 10

7, -7, -10, ...
$\therefore$ The fourth term is even term, which is (-1) times (-10). $\therefore$ 10
Also, the fifth term is odd term, which is 3 less than (10). $\therefore$ 7
Therefore, it goes as, 7, (-)7, (-)10, 10, 7, (-)7, ... , which repeats every four.
$\therefore 47 \div 4$ gives remainder of 3.
$\therefore$ The third term, (-)10 is the answer.

Ans. F

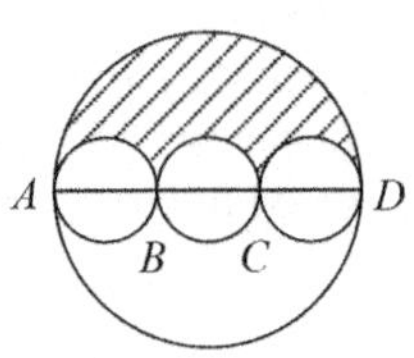

51. AB, BC, CD and AD are diameters of the four circles shown above. If $AB = BC = CD = 2$, what is the area of the shaded region?

A. 12π **B.** 6π **C.** $\frac{9}{2}\pi$ **D.** 3π **E.** 2π

Let R be the radius of the big circle, then $R = 3$.
Also, let r be the radius of the smaller circles, then $r = 1$.
$\therefore$ the area of the shaded region,
$A = \frac{1}{2}(\pi \times 3^2) - \frac{1}{2} \times 3 \times (\pi \times 1^2) = \frac{9\pi}{2} - \frac{3\pi}{2} = \frac{6\pi}{2} = 3\pi$

Ans. D

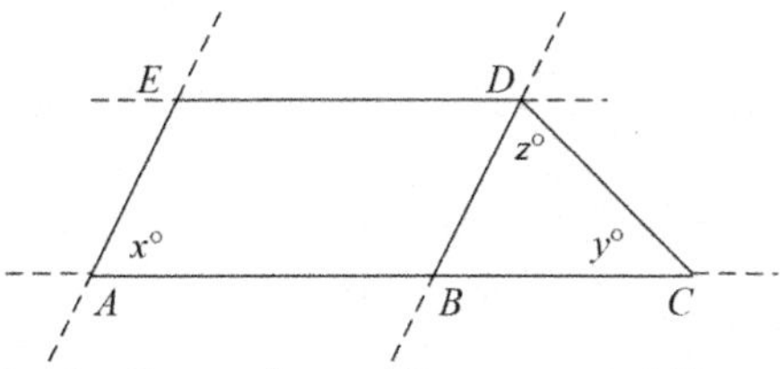

52. In the figure above, if $x + y + z < 180$ and AC is parallel to ED, which of the following must be true?

F. Line AE and BD will intersect when extended.
G. Line BD is not perpendicular to line DC.
H. $x = z$ **J.** $y > x$ **K.** $y < x$

Since $\triangle BDC$ has $\angle B + y + z = 180$ $\therefore \angle B = 180 - y - z$
But, $x + y + z < 180$
$\therefore x < 180 - y - z$ $\therefore x < \angle B$
$\therefore$ Lines AE and BD will intersect, when they are extended.

Ans. F

53. When the radius of a given circle is lengthened by 1 inch, the area is increased by 9π square inches. What is the length, in inches, of the radius of the original circle?

A. 4 **B.** 6 **C.** 8 **D.** 9 **E.** 10

Let r = radius, and A = area of the original circle.
Then, $A + 9\pi = \pi(r + 1)^2$. But $A = \pi r^2$.
$\therefore \pi r^2 + 9\pi = \pi r^2 + 2\pi r + \pi$ $\therefore 2\pi r = 8\pi$ $\therefore r = 4$

Ans. A

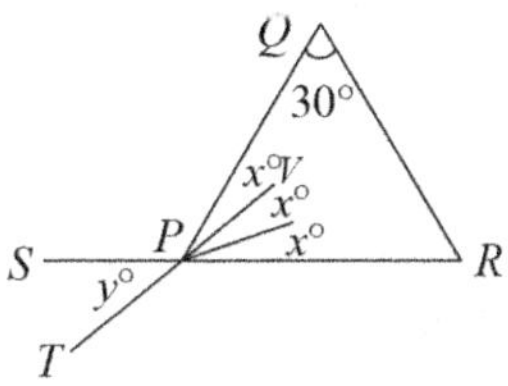

54. In the figure above, $\triangle PQR$ is an isosceles triangle with the vertex angle $\theta = 30°$, and SR and TV are line segments. What is the value of $x + y$?

F. 60 **G.** 75 **H.** 20 **J.** 45 **K.** 15

Since $\triangle PQR$ is an isosceles $\triangle$ with vertex angle of 30°, we get $\angle P = \angle R = 75°$.
$\therefore x = 25°$, and $y = 2x = 50°$
$\therefore x + y = 75°$

Ans. G

55. If A, B, C, D and F are 5 distinct points on the circumference of a circle, how many different chords can be drawn using any 2 of the 5 points?

A. 6 **B.** 8 **C.** 10 **D.** 12 **E.** 24

We are choosing 2 out of 5 points.

Therefore, ${}_5C_2 = \dfrac{5!}{(5-2)!2!} = 10$

Ans. C

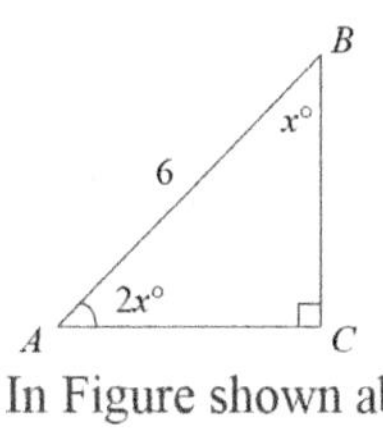

56. In Figure shown above, what is the perimeter?

F. $9 + 3\sqrt{3}$ **G.** 10 **H.** $5 + 2\sqrt{3}$
J. $10 + \sqrt{3}$ **K.** $12 + \sqrt{2}$

Since $2x + x = 3x = 90°$, we have $x = 30°$
Therefore, it is 30°-60°-90° with its sides 3-$3\sqrt{3}$-6 respectively.
So, $P = 3 + 6 + 3\sqrt{3} = 9 + 3\sqrt{3}$

Ans. F

57. If $x = \sqrt[3]{t}$ and $y = t^2$, then $\dfrac{y}{x^5} =$

A. $t^{-\frac{1}{3}}$ **B.** $t^{\frac{1}{3}}$ **C.** $t^{\frac{5}{6}}$ **D.** $t^{\frac{6}{5}}$ **E.** $t^{\frac{10}{3}}$

$x = t^{\frac{1}{3}}, y = t^2$, therefore, $\dfrac{y}{x^5} = \dfrac{t^2}{t^{\frac{1}{3}\cdot 5}} = t^{2-\frac{5}{3}} = t^{\frac{1}{3}}$

Ans. B

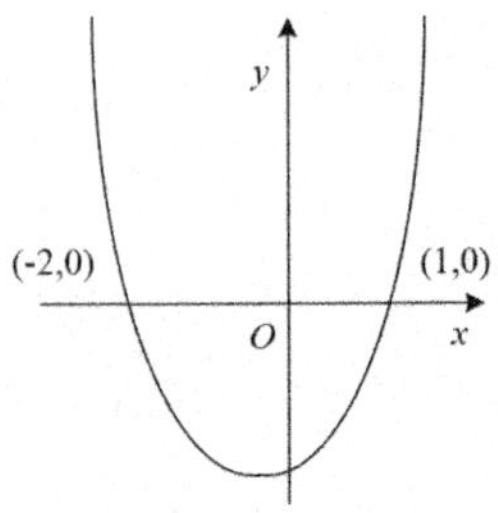

58. Which of the following could be the equation of the parabola in the figure shown above?

F. $y = (x - 2)(x - 3)$
G. $y = (x + 2)(x + 3)$
H. $y = (x + 2)(x - 3)$
J. $y = (x + 2)(x - 1)$
K. $x = (y + 2)(y - 3)$

Since $x = -2$ and $x = 1$ are zeros of the function, we have $(x + 2)(x - 1)$ factors for the function.

Ans. J

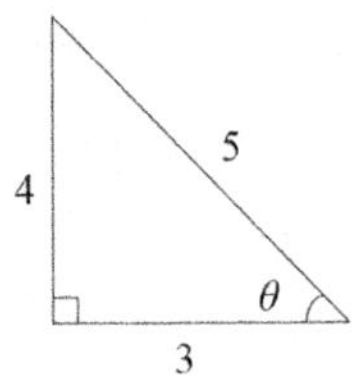

59. In above figure, $\frac{1}{\tan\theta} + \sec\theta =$

A. 0.75 **B.** 1.20 **C.** 1.43 **D.** 2.74 **E.** 2.42

For the given triangle, $\tan\theta = \frac{4}{3}$ and $\cos\theta = \frac{3}{5}$.

Therefore, $\frac{1}{\tan\theta} = \frac{3}{4}$ and $\frac{1}{\cos\theta} = \sec\theta = \frac{5}{3}$.

So, $\frac{1}{\tan\theta} + \sec\theta = \frac{3}{4} + \frac{5}{3} = \frac{9+20}{12} = \frac{29}{12} = 2.42$.

Ans. E

60. For what value of x will $f(x) = 3x^2 - 4x + 5$ have the minimum value?

F. -2 **G.** $-\frac{1}{3}$ **H.** 0 **J.** $\frac{2}{3}$ **K.** 1

The minimum occurs at vertex point.

Since $x = \frac{-b}{2a}$, where $a = 3$, $b = -4$, $c = 5$,

We get $x = \frac{-(-4)}{2(3)} = \frac{4}{6} = \frac{2}{3}$ is the x-coordinates of the vertex point.

Ans. J

Model Test No. 06

60 Questions / 60 Minutes

Directions: Solve each problem, choose the correct answer, and then fill in the corresponding oval on your answer document.

Do not linger over problems that take too much time. Solve as many as you can; then return to the others in the time you have left for this test.

You are permitted to use a calculator on this test. You many use your calculator for any problems you choose, but some of the problems may best be done without using a calculator.

Note: Unless otherwise stated, all of the following should be assumed.

1. Illustrative figures are NOT necessarily drawn to scale.
2. Geometric figures lie in a plane.
3. The word line indicates a straight line.
4. The word average indicates arithmetic mean.

Answer Sheet
Model Test No. 06

1 Ⓐ Ⓑ Ⓒ Ⓓ Ⓔ	16 Ⓕ Ⓖ Ⓗ Ⓙ Ⓚ	31 Ⓐ Ⓑ Ⓒ Ⓓ Ⓔ	46 Ⓕ Ⓖ Ⓗ Ⓙ Ⓚ
2 Ⓕ Ⓖ Ⓗ Ⓙ Ⓚ	17 Ⓐ Ⓑ Ⓒ Ⓓ Ⓔ	32 Ⓕ Ⓖ Ⓗ Ⓙ Ⓚ	47 Ⓐ Ⓑ Ⓒ Ⓓ Ⓔ
3 Ⓐ Ⓑ Ⓒ Ⓓ Ⓔ	18 Ⓕ Ⓖ Ⓗ Ⓙ Ⓚ	33 Ⓐ Ⓑ Ⓒ Ⓓ Ⓔ	48 Ⓕ Ⓖ Ⓗ Ⓙ Ⓚ
4 Ⓕ Ⓖ Ⓗ Ⓙ Ⓚ	19 Ⓐ Ⓑ Ⓒ Ⓓ Ⓔ	34 Ⓕ Ⓖ Ⓗ Ⓙ Ⓚ	49 Ⓐ Ⓑ Ⓒ Ⓓ Ⓔ
5 Ⓐ Ⓑ Ⓒ Ⓓ Ⓔ	20 Ⓕ Ⓖ Ⓗ Ⓙ Ⓚ	35 Ⓐ Ⓑ Ⓒ Ⓓ Ⓔ	50 Ⓕ Ⓖ Ⓗ Ⓙ Ⓚ
6 Ⓕ Ⓖ Ⓗ Ⓙ Ⓚ	21 Ⓐ Ⓑ Ⓒ Ⓓ Ⓔ	36 Ⓕ Ⓖ Ⓗ Ⓙ Ⓚ	51 Ⓐ Ⓑ Ⓒ Ⓓ Ⓔ
7 Ⓐ Ⓑ Ⓒ Ⓓ Ⓔ	22 Ⓕ Ⓖ Ⓗ Ⓙ Ⓚ	37 Ⓐ Ⓑ Ⓒ Ⓓ Ⓔ	52 Ⓕ Ⓖ Ⓗ Ⓙ Ⓚ
8 Ⓕ Ⓖ Ⓗ Ⓙ Ⓚ	23 Ⓐ Ⓑ Ⓒ Ⓓ Ⓔ	38 Ⓕ Ⓖ Ⓗ Ⓙ Ⓚ	53 Ⓐ Ⓑ Ⓒ Ⓓ Ⓔ
9 Ⓐ Ⓑ Ⓒ Ⓓ Ⓔ	24 Ⓕ Ⓖ Ⓗ Ⓙ Ⓚ	39 Ⓐ Ⓑ Ⓒ Ⓓ Ⓔ	54 Ⓕ Ⓖ Ⓗ Ⓙ Ⓚ
10 Ⓕ Ⓖ Ⓗ Ⓙ Ⓚ	25 Ⓐ Ⓑ Ⓒ Ⓓ Ⓔ	40 Ⓕ Ⓖ Ⓗ Ⓙ Ⓚ	55 Ⓐ Ⓑ Ⓒ Ⓓ Ⓔ
11 Ⓐ Ⓑ Ⓒ Ⓓ Ⓔ	26 Ⓕ Ⓖ Ⓗ Ⓙ Ⓚ	41 Ⓐ Ⓑ Ⓒ Ⓓ Ⓔ	56 Ⓕ Ⓖ Ⓗ Ⓙ Ⓚ
12 Ⓕ Ⓖ Ⓗ Ⓙ Ⓚ	27 Ⓐ Ⓑ Ⓒ Ⓓ Ⓔ	42 Ⓕ Ⓖ Ⓗ Ⓙ Ⓚ	57 Ⓐ Ⓑ Ⓒ Ⓓ Ⓔ
13 Ⓐ Ⓑ Ⓒ Ⓓ Ⓔ	28 Ⓕ Ⓖ Ⓗ Ⓙ Ⓚ	43 Ⓐ Ⓑ Ⓒ Ⓓ Ⓔ	58 Ⓕ Ⓖ Ⓗ Ⓙ Ⓚ
14 Ⓕ Ⓖ Ⓗ Ⓙ Ⓚ	29 Ⓐ Ⓑ Ⓒ Ⓓ Ⓔ	44 Ⓕ Ⓖ Ⓗ Ⓙ Ⓚ	59 Ⓐ Ⓑ Ⓒ Ⓓ Ⓔ
15 Ⓐ Ⓑ Ⓒ Ⓓ Ⓔ	30 Ⓕ Ⓖ Ⓗ Ⓙ Ⓚ	45 Ⓐ Ⓑ Ⓒ Ⓓ Ⓔ	60 Ⓕ Ⓖ Ⓗ Ⓙ Ⓚ

MATHEMATICS TEST

60 Minutes – 60 Questions

DIRECTIONS: Solve each problem, choose the correct answer, and then fill in the corresponding oval on your answer document.

Do not linger over problems that take too much time. Solve as many as you can; then return to the others in the time you have left for this test.

You are permitted to use a calculator on this test. You many use your calculator for any problems you choose, but some of the problems may best be done without using a calculator.

Note: Unless otherwise stated, all of the following should be assumed.

1. Illustrative figures are NOT necessarily drawn to scale.
2. Geometric figures lie in a plane.
3. The word line indicates a straight line.
4. The word average indicates arithmetic mean.

---

1. If p is 5 less than triple of q, what is q in terms of p?

A. $15p$
B. $3(p-5)$
C. $3(5-p)$
D. $\frac{p-5}{3}$
E. $\frac{p+5}{3}$

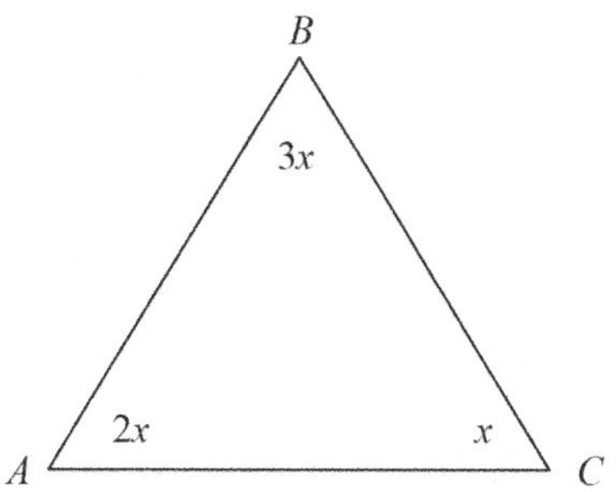

Note: Figure not drawn to scale.

2. In the triangle above, $AC = 10$. What is the length of $\overline{AB}$?

F. 4
G. 5
H. 6
J. 7
K. 8

$$y = 3x - \frac{1}{2}k$$

3. In the equation above, k is a constant. If $y = 5$ when $x = 2$, what is the value of x when $y = 2$?

A. -3
B. -1
C. 1
D. 3
E. 8

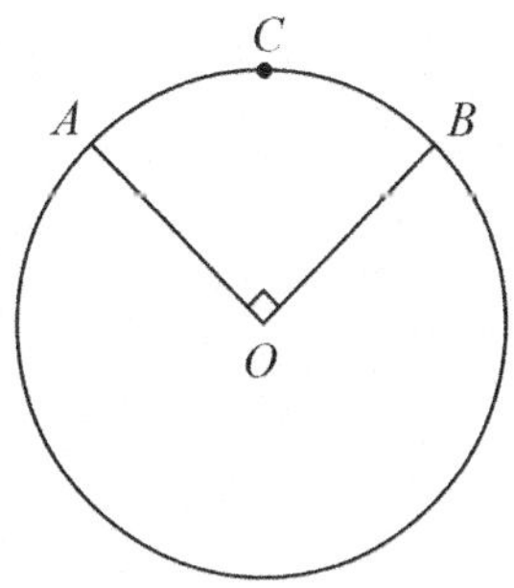

4. In the figure above, the length of arc ACB is 3π, with angle $\angle AOB = 90°$. What is the length of radius of the circle?

F. $\frac{2}{3}$
G. $\frac{3}{2}$
H. 6
J. 2
K. $\frac{7}{2}$

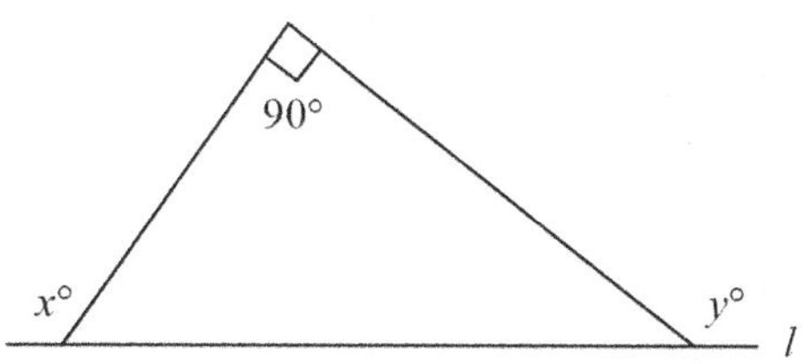

5. In the figure above, what is y in terms of x?

A. $90 - x$
B. $90 + x$
C. $180 - x$
D. $180 + x$
E. $270 - x$

GO ON TO THE NEXT PAGE.

6. In 2001 it was predicted that in the year 2020 the average college tuition in public institution will be approximately \$28,000 per year. This represents a 15 to 20 percent increase from the 2001 figure. Which of the following could have been the average tuition in 2001?

F. \$20,000
G. \$21,000
H. \$22,000
J. \$24,000
K. \$25,000

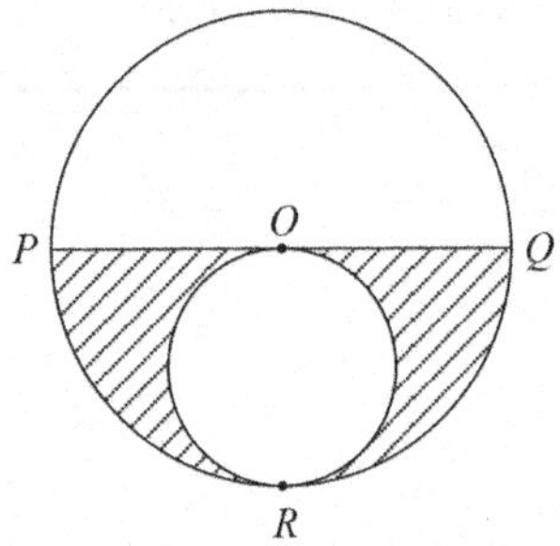

7. In the figure above, O is the center of the larger circle. The smaller circle is tangent to the larger circle at R and contains point O. If the length of diameter PQ is 8, what is the area of the shaded part?

A. 4π
B. 6π
C. 10π
D. 12π
E. 16π

8. When the positive integer x is divided by 3, the remainder is 1. What is the remainder when $x + 15$ is divided by 5?

F. 0
G. 1
H. 2
J. 3
K. 4

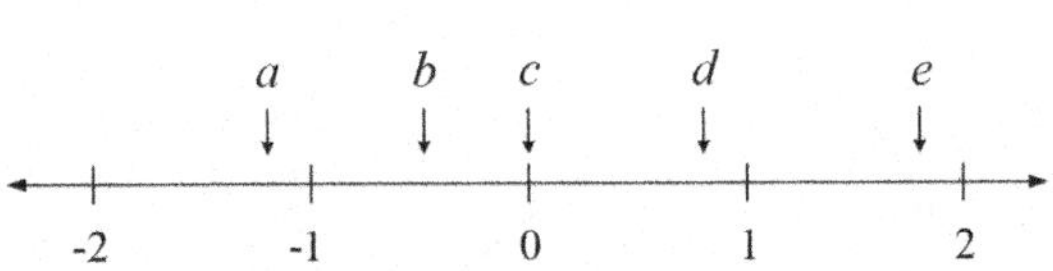

9. The numbers a, b, c, d and e are indicated on the number line above. Of the following, which quantity has the value that is less than -2?

A. a
B. ab
C. ac
D. ad
E. ae

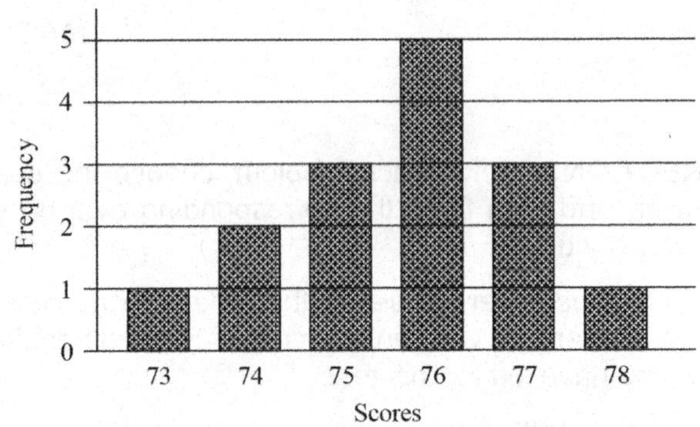

10. The graph above shows the scores of math quiz in Algebra II class. What score is the mode of this group?

F. 73
G. 74
H. 75
J. 76
K. 77

11. For all numbers x, let the function f be defined as $f(x) = x + 3$. Which of the following is equal to $f(2) + f(5)$?

A. $f(7)$
B. $f(9)$
C. $f(10)$
D. $f(16)$
E. $f(22)$

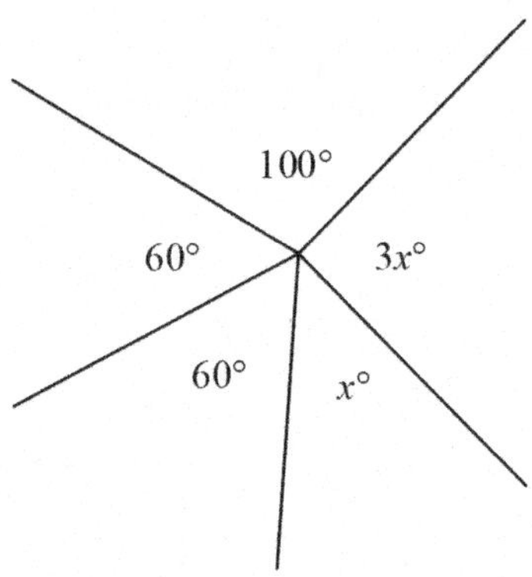

12. What is the value of x in the figure above?

F. 25
G. 35
H. 40
J. 45
K. 50

GO ON TO THE NEXT PAGE.

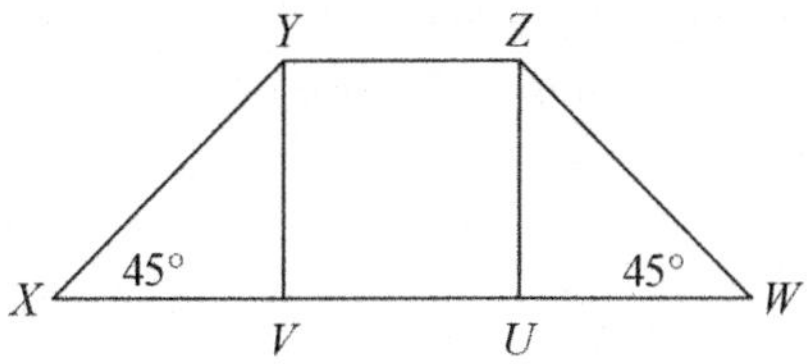

Note: Figure not drawn to scale.

13. In the figure above, the area of square $YZUV$ is 36. What is the perimeter of trapezoid $XYZW$?

A. 20
B. $20 + 6\sqrt{2}$
C. $24 + 12\sqrt{2}$
D. 36
E. 48

14. $15x - ax = 7x$ and $x \neq 0$, what is the value of a?

F. 6
G. 8
H. 10
J. 14
K. 20

15. Lines l, m, r, and s are all different lines that lie in the same plane. If $l \perp m$, $m \perp r$, and $r \perp s$, which of the following must be true?

I. $l \parallel s$
II. $l \parallel r$
III. $m \parallel s$

A. I only
B. I and II only
C. I and III only
D. II and III only
E. I, II, and III

16. One leg of an isosceles right triangle with hypotenuse of length 2 is the hypotenuse of another isosceles right triangle. What is the length of a leg of the second triangle?

F. 1
G. $\sqrt{2}$
H. $\sqrt{3}$
J. $\frac{1}{\sqrt{2}}$
K. $\frac{1}{\sqrt{3}}$

AVERAGE TIME PER WEEK SPENT ON COMMUNITY SERVICE

Grade	Number of Hours
9	-
10	$7\frac{2}{3}$
11	$8\frac{1}{2}$
12	$7\frac{1}{3}$

17. The table above shows the average (arithmetic mean) time a student spent on community service per week during each of 4 years of high school. If the student spent 1,100 hours on community service during 4 years of high school, how many hours did he spend in his 9th grade?
(Assume that the student spent 5 days per week, and each school year contains 180 days.)

A. 306
B. 276
C. 254
D. 216
E. 198

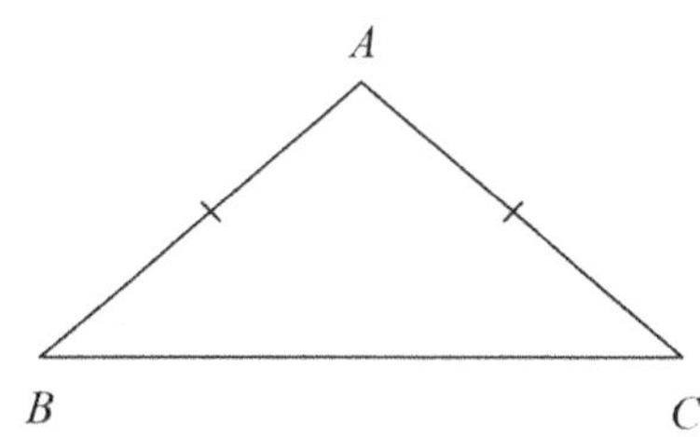

Note: Figure not drawn to scale.

18. In the triangle above, $AB = AC = 13$ and $BC = 24$. What is the area of the triangle?

F. 30
G. 48
H. 60
J. 120
K. 156

19. The sum of the numbers $2x - 3$, $5x + y$, and $-9x + 2y + 3$ is subtracted from 11. In terms of x and y, what is the result?

A. $11 - 4x - 3y$
B. $11 - 4x - y$
C. $11 + 2x - 3y$
D. $19 - 4x - y$
E. $19 + 2x - 3y$

GO ON TO THE NEXT PAGE.

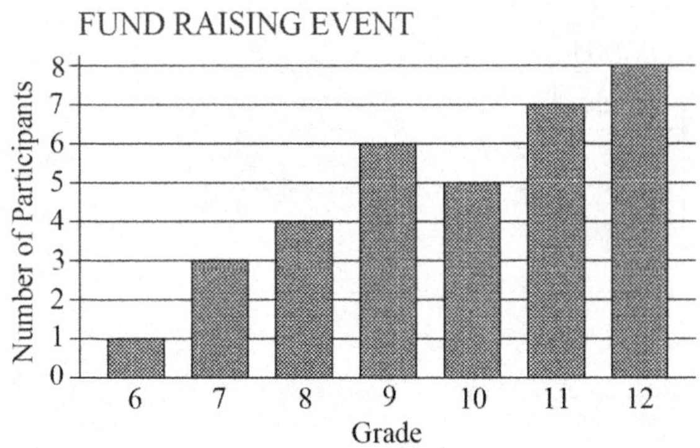

20. The graph above shows the number of participants in a fund-raising event for grades 6 through 12 at Ashton Academy. According to the graph, how many more students in grades 10 through 12 participated than students in grades 6 through 9?

F. 4
G. 6
H. 7
J. 8
K. 11

21. If three dices are rolled, what is the probability that the sum of the numbers on the top faces is 16 or 17?

A. $\frac{1}{48}$
B. $\frac{1}{42}$
C. $\frac{1}{36}$
D. $\frac{1}{24}$
E. $\frac{1}{18}$

22. If $-1 < x < 1$, which of the following is NOT between 0 and 1?

F. $\sqrt{1 - x^2}$
G. $\frac{x}{2} + \frac{1}{2}$
H. $-x^2 + 1$
J. $\sqrt{\frac{x+1}{2}}$
K. $\frac{1}{x}$

23. If the ratio of two positive integers is 3 to 4, which of the following cannot be the sum of the two integers?

A. 15
B. 14
C. 21
D. 28
E. 49

24. $\frac{2a - b}{a + b} = \frac{4}{3}$, what is the value of $\frac{b}{a}$?

F. $-\frac{7}{2}$
G. -1
H. $\frac{7}{2}$
J. 1
K. $\frac{2}{7}$

25. The Cafeteria Restaurant offers 4 different appetizers, 7 different main courses, and 5 different desserts for Dinner Specials. If each order includes exactly one kind of appetizer, one kind of main course and one kind of dessert, how many different kinds of Dinner Specials can be made?

A. 16
B. 48
C. 55
D. 63
E. 140

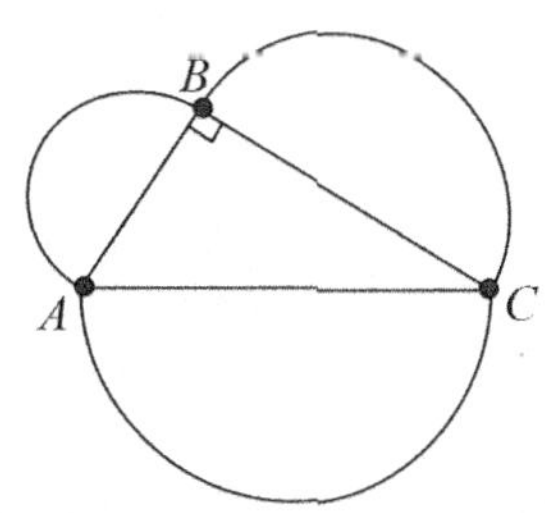

26. There are three semicircles on the sides of right triangle, ΔABC. The area, in square units, of the smallest semicircle is $\frac{9}{2}\pi$, and that of the largest semicircle is $\frac{25}{2}\pi$. What is the perimeter of ΔABC?

F. 18 units
G. 20 units
H. 24 units
J. 30 units
K. $38\frac{1}{2}$ units

GO ON TO THE NEXT PAGE.

27. There are x ounces in y gallons. Which of the following is an expression for the number of ounces in z galleons?

A. $\frac{xz}{y}$

B. $\frac{yz}{x}$

C. $\frac{y}{xz}$

D. $\frac{x}{yz}$

E. xyz

1, 0, -1, ...

28. In a sequence, the first term is 1, and from the second term, any term is defined as 1 less than the proceeding term. What is the sum of the 31st terms in this sequence?

F. 224
G. 15
H. 1
J. 0
K. -434

$2x + 2y + z = 4$
$3x + 3y + 2z = 6$

29. For the system of equations above, what is the value of $x + y$?

A. 1
B. 1.4
C. 2
D. 2.5
E. 3

Set $A = \{1, 3, 5\}$
Set $B = \{0, 2, 4\}$

30. If a and b are to be selected at random from sets A and B, respectively, what is the probability that ab will be a member of set B?

F. $\frac{1}{3}$

G. $\frac{4}{9}$

H. $\frac{1}{2}$

J. $\frac{5}{9}$

K. $\frac{2}{3}$

$m = 3Q71$
$n = 28R7$

31. If m and n above are 4-digit numbers, and if digit Q can be either 1 or 5 and digit R can be either 2 or 4, what is the greatest possible value of $m - n$?

A. 720
B. 724
C. 736
D. 744
E. 748

32. A committee is holding an election of officers - president, treasurer, and secretary. There are 4 men and 5 women in the committee. No one can be a candidate for more than one office. How many different combinations of officers are possible, if the president must be a woman?

F. 224
G. 240
H. 248
J. 280
K. 336

33. For all integers s, let $\boxed{x}$ be defined as follows:

$\boxed{x} = 5x^3$ if x is even.
$\boxed{x} = x^2$ if x is odd.

If $\boxed{7} - \boxed{3} = k$, what is the value of k?

A. 36
B. 38
C. 40
D. 42
E. 48

34. How many different committees of 4 people can be made from a group of 7 people?

F. 27
G. 28
H. 30
J. 32
K. 35

35. When the positive integer n is divided by 5, the remainder is 3. What is the remainder when $3n$ is divided by 5?

A. 0
B. 1
C. 2
D. 3
E. 4

GO ON TO THE NEXT PAGE.

36. If $7x - 3 = \frac{3}{7}$, what is the value of $7x + 7$?

F. $\frac{62}{7}$

G. $\frac{71}{7}$

H. $\frac{73}{7}$

J. $\frac{75}{7}$

K. $\frac{82}{7}$

37. In a certain group of 35 men, each man wears either a hat, a pair of sunglasses, or both. If 27 of the men wear a hat and 20 of the men wear a pair of sunglasses, how many of the men wear both?

A. 7
B. 10
C. 12
D. 15
E. 17

38. A city library has books on archaeology, economics, linguistics, philosophy and theology. Of these books, 35 percent are on economics, 25 percent on philosophy, 20 percent on theology, 15 percent on linguistics, and 45 books are on archaeology. If this library has no other books, how many books are in the library?

F. 900
G. 915
H. 925
J. 930
K. 950

39. If x and y are positive, $y = 2x + 3$, and $xy = 5$, what is the value of $\frac{y}{x}$?

A. $\frac{2}{5}$

B. 1

C. $\frac{5}{2}$

D. 5

E. 10

40. For $n \neq 0$, let $\boxed{n}$ be defined as $\frac{1}{n}$. If a and b are integers such that $0 < a < 1 < b$, which of the following expressions must be true?

F. $\boxed{a} + \boxed{b} < 1$

G. $\boxed{a} \times \boxed{b} > 1$

H. $\boxed{b} - \boxed{a} > 0$

J. $\boxed{b} \div \boxed{a} < 1$

K. None of above

41. If the equations $px = -q$ and $mx = s$ have the same solution x, which of the following <u>must</u> be true?

A. $p = m$ and $q = s$
B. $p - s = q - m$
C. $ps = qm$
D. $pq = ms$
E. $ps + qm = 0$

42. In a particular year, if the third Friday of May is the 18th, what date is the fourth Monday of May?

F. 20th
G. 21st
H. 22nd
J. 27th
K. 28th

43. Ken's weight first decreased by 5 percent and then his new weight increased by 10 percent. His final weight was 165 pounds. What was his initial weight?

A. 142.50 lbs.
B. 147.14 lbs.
C. 157.89 lbs.
D. 172.43 lbs.
E. 173.25 lbs.

44. Suppose a positive integer is defined as symmetric, if it reads the same forward as it does backward. For example, 959 and 8228 are symmetric, whereas 1332 is not. Neither the first nor the last digit can be 0. Which of the following integers is a symmetric?

F. 550
G. 2255
H. 2525
J. 2552
K. 5002

GO ON TO THE NEXT PAGE.

45. Since 1980 the price of a gallon of gasoline has increased from \$1.05 to \$4.20. What is the percent increase in the price of gasoline since 1980?

A. 400%
B. 300%
C. 75%
D. 40%
E. 30%

46. If $x + y = 4$, $y + z = 8$ and $x + z = 6$, then $x + y + z =$

F. 6
G. 9
H. 12
J. 18
K. 24

47. How many different pairs of positive integers have a product of 50?

A. Two
B. Three
C. Four
D. Five
E. Six

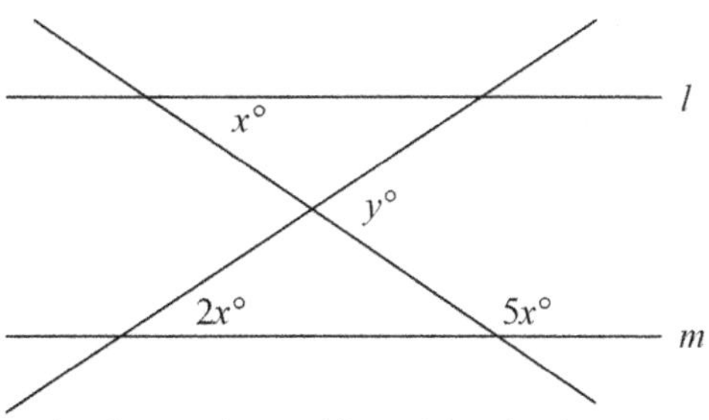

48. In the figure above, $l \parallel m$. What is y?

F. 15
G. 30
H. 45
J. 60
K. 90

49. If the rabbit production in a farm increases in numbers by multiple of two in every 2 months, then how many months have elapsed to have the number of its production reaching 2^{11}?

A. 10
B. 20
C. 30
D. 40
E. 50

a	1	b
3	5	c
4	d	2

50. In the table of numbers above, the sum of the numbers in each row, column, and diagonal is the same. If the letters a, b, c, and d represent the integers 6 through 9, but not necessarily in that order, which letter represents the number 9?

F. a
G. b
H. c
J. d
K. It cannot be determined from the information given.

51. Set X has 6 members and set Y has 7 members. Set Z consists of all members that are in either set X or set Y with the exception of the k common members ($k > 0$). Which of the following represents the number of members in set Z?

A. $13 + k$
B. $13 - k$
C. $13 + 2k$
D. $13 - 2k$
E. 13

52. If $0 < a < b$, then the solution to the equation $\frac{ax - b}{bx - a} = 2$ is $x =$

F. $\frac{2a - b}{2b - a}$
G. $\frac{2a + b}{2b - a}$
H. $\frac{a - b}{b - a}$
J. $\frac{2b - a}{2a - b}$
K. $\frac{2b + a}{2a + b}$

53. If a side of a cube is 1 foot, what is the volume of the cube, in cubic inches?

A. 1
B. 1728
C. 1440
D. 256
E. 625

GO ON TO THE NEXT PAGE.

54. Let the operations $\triangle$ and $\frown$ be defined for all real numbers a and b as follows.

$a \triangle b = 2a + 3b$
$a \frown b = 3a + 2b$

If $3 \triangle (4x) = (3x) \frown 4$, what is the value of x?

F. 1
G. $\frac{2}{3}$
H. $\frac{1}{3}$
J. 2
K. $\frac{1}{2}$

55. If $f(x) = \frac{1}{2}|x| + 3$, for which of the following values of x does $f(x) = f(-x)$?

A. -10 only
B. -10 and 10 only
C. All real x
D. All real x except 10
E. All real x except -10 and 10

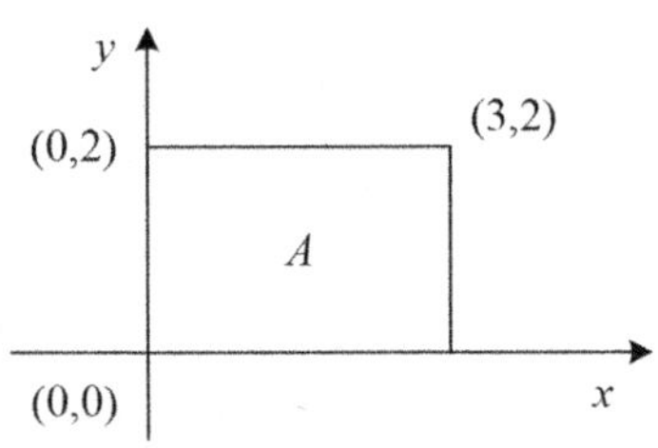

56. In the given figure, rectangle A contains all points (x,y). What is the area of a rectangle that contains all points $(2x, y - 1)$?

F. 12
G. 14
H. 18
J. 24
K. 48

57. If $\sin\theta = m$ and $0 < \theta < 90°$, then $\tan\theta =$

A. $\frac{1}{m^2}$
B. $\frac{m}{\sqrt{1 - m^2}}$
C. $\frac{1 - m^2}{m}$
D. $\frac{m}{1 - m^2}$
E. $\frac{m^2}{\sqrt{1 - m^2}}$

58. If $f(x) = 3\sqrt{x} + 1$, for all $x > 0$, then $f^{-1}(x) =$

F. $\frac{1}{9}(x - 1)^2$
G. $\frac{1}{3}(x + 2)^2$
H. $x^2 + 1$
J. $(x - 4)^2$
K. $(x + 2)^2 + 1$

59. When $2x^2 + 6x + k$ is divided by $x - 2$, the remainder is 2. Which of the following is the value of k?

A. -18
B. -16
C. -10
D. 6
E. 14

60. y varies directly as the square of x. When $y = 6$, $x = 4$. If $y = 27$, then x could equal

F. $-6\sqrt{2}$
G. $-8\sqrt{2}$
H. $-10\sqrt{3}$
J. $-16\sqrt{2}$
K. $-64\sqrt{3}$

END OF TEST 6
STOP! DO NOT TURN THE PAGE UNTIL TOLD TO DO SO.
DO NOT RETURN TO THE PREVIOUS TEST.

Answer Key

Model Test No. 06

1	E	16	F	31	D	46	G
2	G	17	C	32	J	47	B
3	C	18	C	33	C	48	K
4	H	19	C	34	K	49	B
5	E	20	G	35	E	50	J
6	J	21	D	36	H	51	D
7	A	22	K	37	C	52	F
8	K	23	A	38	F	53	B
9	E	24	K	39	D	54	G
10	J	25	E	40	J	55	C
11	C	26	H	41	E	56	F
12	G	27	A	42	K	57	B
13	C	28	K	43	C	58	F
14	G	29	C	44	J	59	A
15	D	30	J	45	B	60	F

How to Score the ACT Mathematics Test

When you take an actual ACT Mathematics Test, your answer sheet will be "read" by a scanning machine that will record your responses to each question. Then a computer will compare your answers with the correct answers and produce your raw score. You get one point for each correct answer. For each wrong answer, there is no penalty. Therefore, you must guess and answer all the questions to get a higher score. This raw score is converted to a scaled score that is reported to you and to the colleges you specify.

Finding Your Scale Test Score

STEP 1: Count the number of right answers.
Enter the total here: ______________________

STEP 2: The number you obtained in Step 1 is your raw score.
Look up the conversion table for ACT-Math, and find your score.

Scaled Score Conversion Table
ACT Mathematics Test

Scale Score	Mathematics
36	60
35	58-59
34	56-57
33	55
32	54
31	52-53
30	50-51
29	48-49
28	46-47
27	43-45
26	41-42
25	39-40
24	37-38
23	35-36
22	33-34
21	31-32
20	30
19	27-29
18	25-26
17	22-24
16	18-21
15	15-17
14	12-14
13	09-11
12	08
11	06-07
10	05
09	04
08	-
07	03
06	02
05	-
04	-
03	01
02	-
01	00

1. If p is 5 less than triple of q, what is q in terms of p?

A. $15p$ **B.** $3(p-5)$ **C.** $3(5-p)$ **D.** $\frac{p-5}{3}$ **E.** $\frac{p+5}{3}$

$p = 3q - 5 \quad \therefore q = \frac{p+5}{3}$

Ans. E

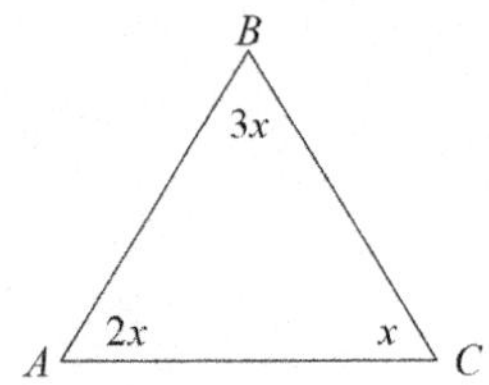

2. In the triangle above, $AC = 10$. What is the length of $\overline{AB}$?

F. 4 **G.** 5 **H.** 6 **J.** 7 **K.** 8

$x + 2x + 3x = 180° \quad \therefore 6x = 180° \quad \therefore x = 30°$
Therefore, the triangle is, 30-60-90 special right triangle with the side length $1:\sqrt{3}:2$, or $5:5\sqrt{3}:10$. $\therefore \overline{AB} = 5$

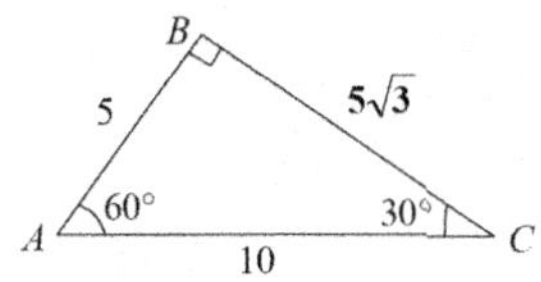

Ans. G

$$y = 3x - \frac{1}{2}k$$

3. In the equation above, k is a constant. If $y = 5$ when $x = 2$, what is the value of x when $y = 2$?

A. -3 **B.** -1 **C.** 1 **D.** 3 **E.** 8

$x = 2, y = 5$

$\therefore 5 = 3(2) - \frac{1}{2}k \quad \therefore k = 2 \quad \therefore$ We get, $y = 3x - 1$

$\therefore$ For a given $y = 2$, we have $2 = 3x - 1$. $\therefore x = 1$

Ans. C

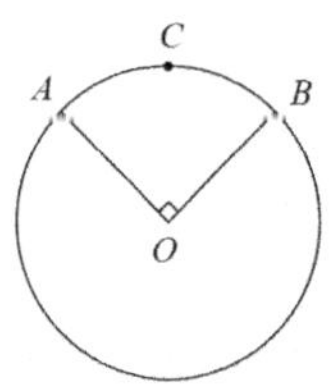

4. In the figure above, the length of arc ACB is 3π, with angle $\angle AOB = 90°$. What is the length of radius of the circle?

F. $\frac{2}{3}$ **G.** $\frac{3}{2}$ **H.** 6 **J.** 2 **K.** $\frac{7}{2}$

arc $ACB = 3\pi = r \cdot \theta = r \cdot (90°) = r\frac{\pi}{2}$

$\therefore r = 6$

Ans. H

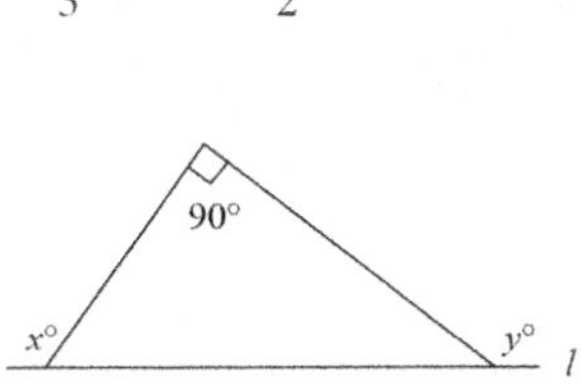

5. In the figure above, what is y in terms of x?

A. $90 - x$ **B.** $90 + x$ **C.** $180 - x$
D. $180 + x$ **E.** $270 - x$

$(180° - x) + (180° - y) = 90°$
$\therefore 360° - x - y = 90°$
$\therefore y = 270° - x$

Ans. E

6. In 2001 it was predicted that in the year 2020 the average college tuition in public institution will be approximately \$28,000 per year. This represents a 15 to 20 percent increase from the 2001 figure. Which of the following could have been the average tuition in 2001?

F. \$20,000 **G.** \$21,000 **H.** \$22,000
J. \$24,000 **K.** \$25,000

Since the percent increase over the years runs from 15% to 20%, we get, $28{,}000 = (1 + 0.15)x = 1.15x$ or $28{,}000 = (1 + 0.20)x = 1.20x$.

$\therefore \dfrac{28{,}000}{1.20} \le x \le \dfrac{28{,}000}{1.15}$ or $23{,}333 \le x \le 24{,}348$

Ans. J

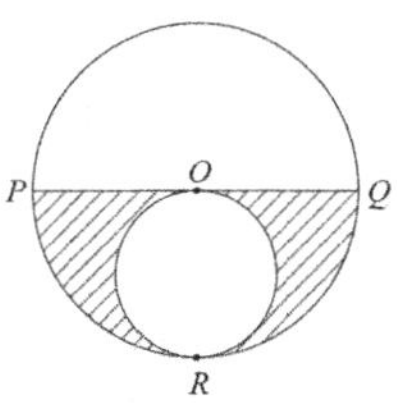

7. In the figure above, O is the center of the larger circle. The smaller circle is tangent to the larger circle at R and contains point O. If the length of diameter PQ is 8, what is the area of the shaded part?

A. 4π **B.** 6π **C.** 10π **D.** 12π **E.** 16π

The area of the shaded part is,

$\frac{1}{2}$(Big Circle) – small circle $= \frac{1}{2}\pi \cdot 4^2 - \pi \cdot 2^2 = 4\pi$

Ans. A

8. When the positive integer x is divided by 3, the remainder is 1. What is the remainder when $x + 15$ is divided by 5?

F. 0 **G.** 1 **H.** 2 **J.** 3 **K.** 4

Let's say $x = 4$, then, $x + 15 = 19$.

$\therefore 19 \div 5 = 3$ with $R = 4$

Ans. K

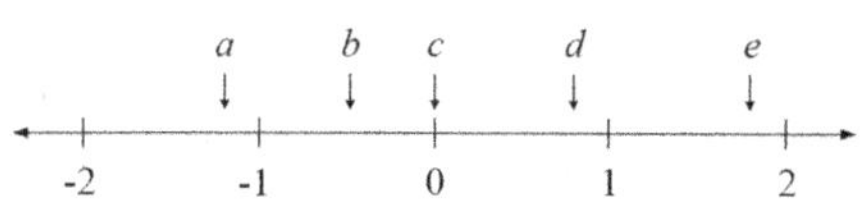

9. The numbers a, b, c, d and e are indicated on the number line above. Of the following, which quantity has the value that is less than -2?

A. a **B.** ab **C.** ac **D.** ad **E.** ae

Let $a \approx -1.2$, $b \approx -0.5$, $c \approx 0$, $d \approx 0.8$ and $e \approx 1.8$.

Then, $ae = (-1.2)(1.8) = -2.16$.

Ans. E

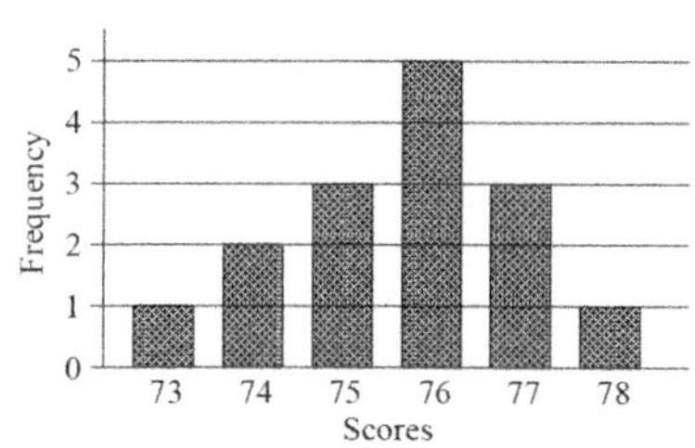

10. The graph above shows the scores of math quiz in Algebra II class. What score is the mode of this group?

F. 73 **G.** 74 **H.** 75 **J.** 76 **K.** 77

The mode is the number that is most frequently occurring in data set.

$\therefore$ In this case, it will be 76, which has its frequency 5.

Ans. J

11. For all numbers x, let the function f be defined as $f(x) = x + 3$. Which of the following is equal to $f(2) + f(5)$?

A. $f(7)$ **B.** $f(9)$ **C.** $f(10)$ **D.** $f(16)$ **E.** $f(22)$

$f(2) = 2 + 3 = 5$, $f(5) = 5 + 3 = 8$

$\therefore f(2) + f(5) = 5 + 8 = 13$, which is, $f(10) = 10 + 3 = 13$.

Ans. C

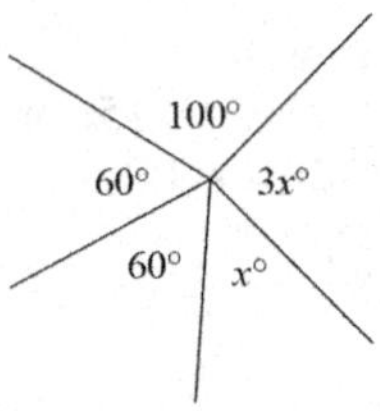

12. What is the value of x in the figure above?

F. 25 **G.** 35 **H.** 40 **J.** 45 **K.** 50

$4x = 360 - 100 - 60 - 60 = 140$

$\therefore x = \frac{140}{4} = 35$

Ans. G

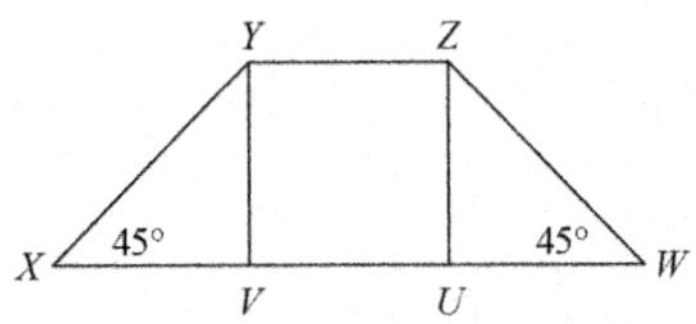

13. In the figure above, the area of square $YZUV$ is 36. What is the perimeter of trapezoid $XYZW$?

A. 20 **B.** $20 + 6\sqrt{2}$ **C.** $24 + 12\sqrt{2}$
D. 36 **E.** 48

Since □$YZUV$ is a SQUARE with its area 36, it has side length of 6 each.
∴ From its figure, we get ΔXYV,

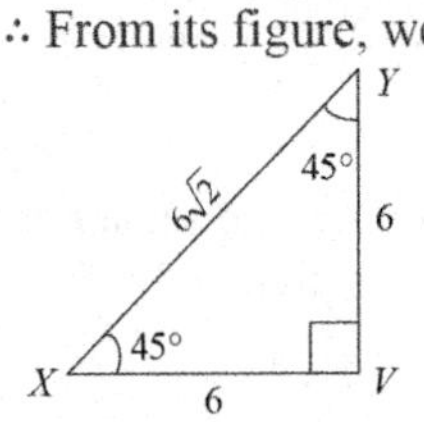

$\therefore$ perimeter $P = 2(6\sqrt{2}) + (6 + 6 + 6 + 6) = 24 + 12\sqrt{2}$

Ans. C

14. $15x - ax = 7x$ and $x \neq 0$, what is the value of a?

F. 6 **G.** 8 **H.** 10 **J.** 14 **K.** 20

$15x - ax = 7x$

$\therefore 15x - 7x = ax$, or $8x = ax$ $\therefore a = 8$

Ans. G

15. Lines l, m, r, and s are all different lines that lie in the same plane. If $l \perp m$, $m \perp r$, and $r \perp s$, which of the following must be true?

I. $l \parallel s$
II. $l \parallel r$
III. $m \parallel s$

A. I only **B.** I and II only **C.** I and III only
D. II and III only **E.** I, II, and III

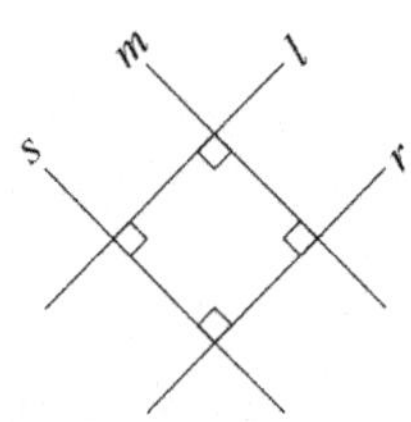

$\therefore s \parallel m$ and $l \parallel r$

Ans. D

16. One leg of an isosceles right triangle with hypotenuse of length 2 is the hypotenuse of another isosceles right triangle. What is the length of a leg of the second triangle?

F. 1 **G.** $\sqrt{2}$ **H.** $\sqrt{3}$ **J.** $\frac{1}{\sqrt{2}}$ **K.** $\frac{1}{\sqrt{3}}$

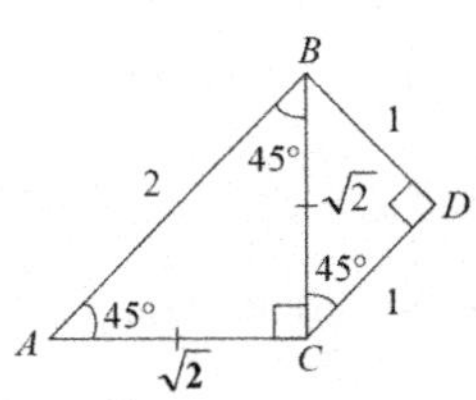

Ans. F

AVERAGE TIME PER WEEK SPENT ON COMMUNITY SERVICE

Grade	Number of Hours
9	-
10	$7\frac{2}{3}$
11	$8\frac{1}{2}$
12	$7\frac{1}{3}$

17. The table above shows the average (arithmetic mean) time a student spent on community service per week during each of 4 years of high school. If the student spent 1,100 hours on community service during 4 years of high school, how many hours did he spend in his 9th grade?
(Assume that the student spent 5 days per week, and each school year contains 180 days.)

A. 306 **B.** 276 **C.** 254 **D.** 216 **E.** 198

Since the school year contains 180 days with 5 days per week, we get $180 \div 5 = 36$ weeks/year.

$\therefore$ For 10th, $7\frac{2}{3} \times 36$ wks $= 276$

For 11th, $8\frac{1}{2} \times 36$ wks $= 306$

For 12th, $7\frac{1}{3} \times 36$ wks $= 264$

$\therefore$ For 9th, we get $1{,}100 - (276 + 306 + 264) = 254$

Ans. C

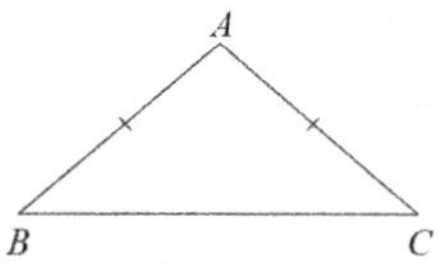

18. In the triangle above, $AB = AC = 13$ and $BC = 24$. What is the area of the triangle?

F. 30 **G.** 48 **H.** 60 **J.** 120 **K.** 156

Since the triangle has base 24, height 5,

we get area $\frac{1}{2}(24)(5) = 60$

Ans. C

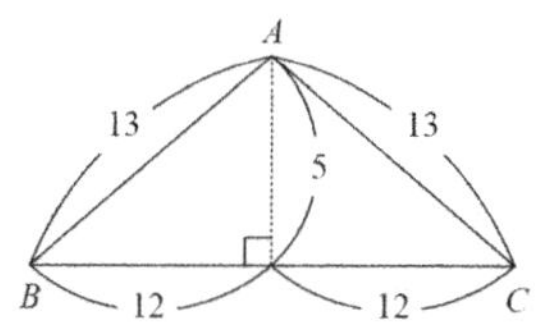

19. The sum of the numbers $2x - 3$, $5x + y$, and $-9x + 2y + 3$ is subtracted from 11. In terms of x and y, what is the result?

A. $11 - 4x - 3y$ **B.** $11 - 4x - y$ **C.** $11 + 2x - 3y$
D. $19 - 4x - y$ **E.** $19 + 2x - 3y$

$11 - [(2x - 3) + (5x + y) + (-9x + 2y + 3)] =$
$11 - (-2x + 3y) = 2x - 3y + 11$

Ans. C

FUND RAISING EVENT

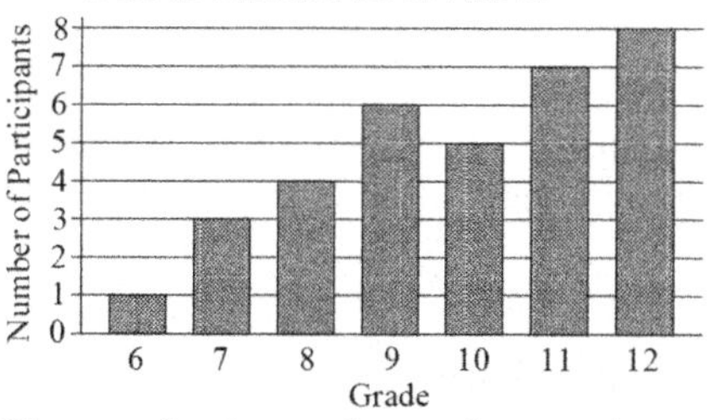

20. The graph above shows the number of participants in a fund-raising event for grades 6 through 12 at Ashton Academy. According to the graph, how many more students in grades 10 through 12 participated than students in grades 6 through 9?

F. 4 **G.** 6 **H.** 7 **J.** 8 **K.** 11

The number of students participated, for $6 - 9^{th}$ grade, $x = 1 + 3 + 4 + 6 = 14$.
Also, for $10 - 12^{th}$ grade, $y = 5 + 7 + 8 = 20$.
$\therefore 20 - 14 = 6$

Ans. G

21. If three dices are rolled, what is the probability that the sum of the numbers on the top faces is 16 or 17?

A. $\frac{1}{48}$ **B.** $\frac{1}{42}$ **C.** $\frac{1}{36}$ **D.** $\frac{1}{24}$ **E.** $\frac{1}{18}$

The possible outcomes for each dice are one through six.
∴ To get 16, we have 6 - 5 - 5, 6 - 6 - 4,
and also, to get 17, 6 - 6 - 5 only. ∴ 3 cases.
But since each case has 3 ways of positioning,
such as 6 - 5 - 5 or 5 - 6 - 5 or 5 - 5 - 6, … etc.,
we get $p(16 \text{ or } 17) = \frac{3 \times 3}{6 \times 6 \times 6} = \frac{1}{24}$.

Ans. D

22. If $-1 < x < 1$, which of the following is NOT between 0 and 1?

F. $\sqrt{1 - x^2}$ **G.** $\frac{x}{2} + \frac{1}{2}$ **H.** $-x^2 + 1$ **J.** $\sqrt{\frac{x+1}{2}}$ **K.** $\frac{1}{x}$

Let $x = \frac{1}{2}$, then all of the answer choices satisfy,
except **K.** $\frac{1}{x} = \frac{1}{\frac{1}{2}} = 2 > 1$.

Ans. K

23. If the ratio of two positive integers is 3 to 4, which of the following cannot be the sum of the two integers?

A. 15 **B.** 14 **C.** 21 **D.** 28 **E.** 49

Let two positive integers be $3k$ and $4k$, then the sum is,
$(3k + 4k) = 7k$. ∴ multiple of 7's.

Ans. A

24. $\frac{2a - b}{a + b} = \frac{4}{3}$, what is the value of $\frac{b}{a}$?

F. $-\frac{7}{2}$ **G.** -1 **H.** $\frac{7}{2}$ **J.** 1 **K.** $\frac{2}{7}$

By cross multiplication, $3(2a - b) = 4(a + b)$
$\therefore 6a - 3b = 4a + 4b \quad \therefore 2a = 7b \quad \therefore \frac{b}{a} = \frac{2}{7}$

Ans. K

25. The Cafeteria Restaurant offers 4 different appetizers, 7 different main courses, and 5 different desserts for Dinner Specials. If each order includes exactly one kind of appetizer, one kind of main course and one kind of dessert, how many different kinds of Dinner Specials can be made?

A. 16 **B.** 48 **C.** 55 **D.** 63 **E.** 140

The number of different combination for the Dinner Special is,
${}_4C_1 \times {}_7C_1 \times {}_5C_1 = 4 \times 7 \times 5 = 140$.

Ans. E

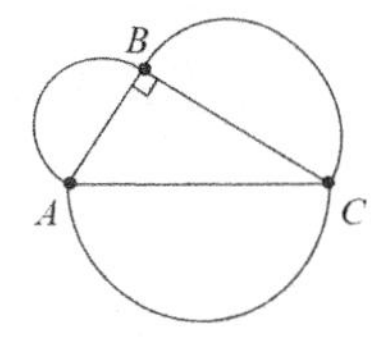

26. There are three semicircles on the sides of right triangle, $\triangle ABC$. The area, in square units, of the smallest semicircle is $\frac{9}{2}\pi$, and that of the largest semicircle is $\frac{25}{2}\pi$. What is the perimeter of $\triangle ABC$?

F. 18 units **G.** 20 units **H.** 24 units

J. 30 units **K.** $38\frac{1}{2}$ units

Let r_1 = radius of the smallest, r_2 = radius of the medium and r_3 = radius of the largest circle.
Then, the area of the smallest is, $\pi r_1^2 = 2\frac{9\pi}{2} = 9\pi$, $\therefore r_1 = 3$
Also, the largest circle is, $\pi r_3^2 = 2\frac{25\pi}{2} = 25\pi$, $\therefore r_3 = 5$
Now, we get the right triangle of 3 – 4 – 5.
Since 3 – 4 – 5 are the radius,
6 – 8 – 10 becomes the diameter.
$\therefore p = 6 + 8 + 10 = 24$

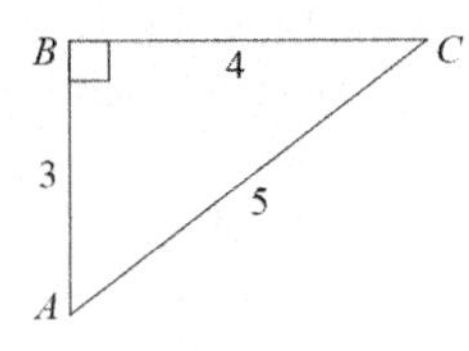

Ans. H

27. There are x ounces in y gallons. Which of the following is an expression for the number of ounces in z galleons?

A. $\frac{xz}{y}$ **B.** $\frac{yz}{x}$ **C.** $\frac{y}{xz}$ **D.** $\frac{x}{yz}$ **E.** xyz

$\frac{x \text{ oz}}{y \text{ gal}} = \frac{?}{z \text{ gal}}$ $\therefore xz = y(?)$ $\therefore ? = \frac{xz}{y}$

Ans. A

1, 0, -1, ...

28. In a sequence, the first term is 1, and from the second term, any term is defined as 1 less than the proceeding term. What is the sum of the 31st terms in this sequence?

F. 224 **G.** 15 **H.** 1 **J.** 0 **K.** -434

The sequence becomes, 1, 0, -1, -2, -3, ... $\therefore a_1 = 1, d = -1$

$\therefore$ Using the formula of the Arithmetic Sequence,

$a_n = a_1 + (n-1)d$, $s_n = \frac{(a_1 + a_n)n}{2}$, we get $a_{31} = a_1 + (31-1)d$

$= 1 + (30)(-1) = -29$ $\therefore s_{31} = \frac{(1-29)(31)}{2} = -434$

Ans. K

$2x + 2y + z = 4$
$3x + 3y + 2z = 6$

29. For the system of equations above, what is the value of $x + y$?

A. 1 **B.** 1.4 **C.** 2 **D.** 2.5 **E.** 3

Let $2x + 2y + z = 4$ be eq ① and $3x + 3y + 2z = 6$ be eq ②.
Then, 2 × eq ① – eq ② =

$$\begin{array}{rl} & 4x + 4y + 2z = 8 \\ - & 3x + 3y + 2z = 6 \\ \hline & x + y \quad\; = 2 \end{array}$$

Ans. C

Set $A = \{1, 3, 5\}$
Set $B = \{0, 2, 4\}$

30. If a and b are to be selected at random from sets A and B, respectively, what is the probability that ab will be a member of set B?

F. $\frac{1}{3}$ **G.** $\frac{4}{9}$ **H.** $\frac{1}{2}$ **J.** $\frac{5}{9}$ **K.** $\frac{2}{3}$

The possible combination of ab for being the member of set B is, (1 × 0), (3 × 0), (5 × 0), (1 × 2), (1 × 4).

$\therefore$ 5 out of (3 × 3) $\therefore p = \frac{5}{9}$

Ans. J

$m = 3Q71$
$n = 28R7$

31. If m and n above are 4-digit numbers, and if digit Q can be either 1 or 5 and digit R can be either 2 or 4, what is the <u>greatest</u> possible value of $m - n$?

A. 720 **B.** 724 **C.** 736 **D.** 744 **E.** 748

To get the maximum of $m - n$, m has to be the greatest possible.
$\therefore m = 3571$ and n has to be the least possible.
$\therefore n = 2827$ $\therefore m - n = 3571 - 2827 = 744$

Ans. D

32. A committee is holding an election of officers - president, treasurer, and secretary. There are 4 men and 5 women in the committee. No one can be a candidate for more than one office. How many different combinations of officers are possible, if the president must be a woman?

F. 224 **G.** 240 **H.** 248 **J.** 280 **K.** 336

There are 4 men and 5 women.
$\therefore$ total = 9. But the president must be women.

$$\begin{array}{ccc} \boxed{P} \;, & \boxed{T} \;, & \boxed{S} \\ \uparrow & \uparrow & \uparrow \\ {}_5C_1 & {}_8C_1 & {}_7C_1 \end{array}$$

Once the president is selected, the rest can be either men or women from the remaining 8.

$\therefore {}_5C_1 \times {}_8C_1 \times {}_7C_1 = 5 \times 8 \times 7 = 280$

Ans. J

33. For all integers s, let $\boxed{x}$ be defined as follows:

$\boxed{x} = 5x^3$ if x is even.

$\boxed{x} = x^2$ if x is odd.

If $\boxed{7} - \boxed{3} = k$, what is the value of k?

A. 36 **B.** 38 **C.** 40 **D.** 42 **E.** 48

$\boxed{7} = 7^2, \boxed{3} = 3^2$

$\therefore 49 - 9 = 40$

Ans. C

34. How many different committees of 4 people can be made from a group of 7 people?

F. 27 **G.** 28 **H.** 30 **J.** 32 **K.** 35

${}_7C_4 = 35$

Ans. K

35. When the positive integer n is divided by 5, the remainder is 3. What is the remainder when $3n$ is divided by 5?

A. 0 **B.** 1 **C.** 2 **D.** 3 **E.** 4

Let $n = 8$. Then $3n = 24$

$\therefore 24 \div 5 = 4$ and $R = 4$

Ans. E

36. If $7x - 3 = \frac{3}{7}$, what is the value of $7x + 7$?

F. $\frac{62}{7}$ **G.** $\frac{71}{7}$ **H.** $\frac{73}{7}$ **J.** $\frac{75}{7}$ **K.** $\frac{82}{7}$

$7x - 3 = \frac{3}{7}$

$\therefore 7x + 7 = (7x - 3) + 10 = \frac{3}{7} + 10 = \frac{73}{7}$

Ans. H

37. In a certain group of 35 men, each man wears either a hat, a pair of sunglasses, or both. If 27 of the men wear a hat and 20 of the men wear a pair of sunglasses, how many of the men wear both?

A. 7 **B.** 10 **C.** 12 **D.** 15 **E.** 17

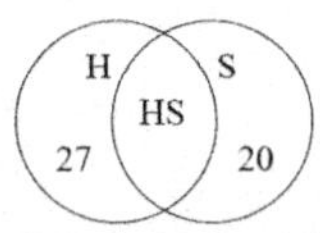

$(H \cup S) = H + S - (H \cap S)$

$\therefore 35 = 27 + 20 - (H \cap S)$

$\therefore (H \cap S) = 27 + 20 - 35 = 12$

Ans. C

38. A city library has books on archaeology, economics, linguistics, philosophy and theology. Of these books, 35 percent are on economics, 25 percent on philosophy, 20 percent on theology, 15 percent on linguistics, and 45 books are on archaeology. If this library has no other books, how many books are in the library?

F. 900 **G.** 915 **H.** 925 **J.** 930 **K.** 950

Let T = the total number of books in the library. Then, since Archeology has its percentage of (100 – 35 – 25 – 20 – 15), which is 5%, and 45 books for that, we get 45 = (0.05)T.

$\therefore T = \frac{45}{0.05} = 900$

Ans. F

39. If x and y are positive, $y = 2x + 3$, and $xy = 5$, what is the value of $\frac{y}{x}$?

A. $\frac{2}{5}$ **B.** 1 **C.** $\frac{5}{2}$ **D.** 5 **E.** 10

$y = 2x + 3$

$\therefore xy = x(2x + 3) = 5$

$\therefore 2x^2 + 3x - 5 = 0$, or $(2x + 5)(x - 1) = 0$

$\therefore x = -\frac{5}{2}$, or $x = 1$

But since x is positive, $x = 1$ only.

$\therefore y = 2x + 3 = 2(1) + 3 = 5$

$\therefore \frac{y}{x} = \frac{5}{1} = 5$

Ans. D

40. For $n \neq 0$, let $\boxed{n}$ be defined as $\frac{1}{n}$. If a and b are integers such that $0 < a < 1 < b$, which of the following expressions must be true?

F. $\boxed{a} + \boxed{b} < 1$
G. $\boxed{a} \times \boxed{b} > 1$
H. $\boxed{b} - \boxed{a} > 0$
J. $\boxed{b} \div \boxed{a} < 1$
K. None of above

Let $a = \frac{1}{2}$, $b = 2$. Then,

F. $\frac{1}{\frac{1}{2}} + \frac{1}{2} = 2 + \frac{1}{2} > 1$ ∴ No. **G.** $\frac{1}{\frac{1}{2}} \times \frac{1}{2} = 1$ ∴ No.

H. $\frac{1}{2} - \frac{1}{\frac{1}{2}} = \frac{1}{2} - 2 = -\frac{3}{2}$ ∴ No. **J.** $\frac{1}{2} \div \frac{1}{\frac{1}{2}} = \frac{1}{2} \div 2 = \frac{1}{4} < 1$ ∴ Yes.

Ans. J

41. If the equations $px = -q$ and $mx = s$ have the same solution x, which of the following must be true?

A. $p = m$ and $q = s$
B. $p - s = q - m$
C. $ps = qm$
D. $pq = ms$
E. $ps + qm = 0$

$px = -q$ ∴ $x = -\frac{q}{p}$

$mx = s$ ∴ $x = \frac{s}{m}$

∴ $-\frac{q}{p} = \frac{s}{m}$

∴ $-qm = ps$

∴ $ps + qm = 0$

Ans. E

42. In a particular year, if the third Friday of May is the 18th, what date is the fourth Monday of May?

F. 20th **G.** 21st **H.** 22nd **J.** 27th **K.** 28th

1,	2,	3,	4,	…, 7,	…, 11,	…, 14,	…, 18,	…, 21,	…, 28
↑	↑	↑	↑	↑	↑	↑	↑	↑	↑
T	W	Th	1F	1M	2F	2M	3F	3M	4M

Ans. K

43. Ken's weight first decreased by 5 percent and then his new weight increased by 10 percent. His final weight was 165 pounds. What was his initial weight?

A. 142.50 lbs. **B.** 147.14 lbs. **C.** 157.89 lbs.
D. 172.43 lbs. **E.** 173.25 lbs.

Let k = Ken's weight. Then, $k(1 - 0.05)(1 + 0.10) = 165$

∴ $k(0.95)(1.10) = 165$

∴ $k = 157.89$

Ans. C

44. Suppose a positive integer is defined as symmetric, if it reads the same forward as it does backward. For example, 959 and 8228 are symmetric, whereas 1332 is not. Neither the first nor the last digit can be 0. Which of the following integers is a symmetric?

F. 550 **G.** 2255 **H.** 2525 **J.** 2552 **K.** 5002

25 | 52 is symmetric.

Ans. J

45. Since 1980 the price of a gallon of gasoline has increased from \$1.05 to \$4.20. What is the percent increase in the price of gasoline since 1980?

A. 400% **B.** 300% **C.** 75% **D.** 40% **E.** 30%

% Inc. $= \frac{4.20 - 1.05}{1.05} = 3$

∴ 300%

Ans. B

46. If $x + y = 4$, $y + z = 8$ and $x + z = 6$, then $x + y + z =$

F. 6 **G.** 9 **H.** 12 **J.** 18 **K.** 24

$(x + y) + (y + z) + (x + z) = 4 + 8 + 6 = 18$

or $2(x + y + z) = 18$

∴ $(x + y + z) = 9$

Ans. G

47. How many different pairs of positive integers have a product of 50?

A. Two **B.** Three **C.** Four **D.** Five **E.** Six

$50 = 2 \times 5 \times 5$
$\therefore 2 \times 25,\ 10 \times 5,\ 1 \times 50$
$\therefore$ Three of them.

Ans. B

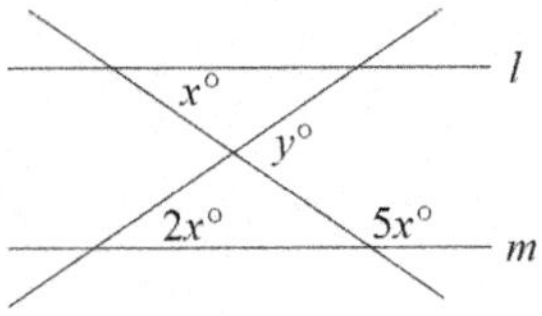

48. In the figure above, $l \parallel m$. What is y?

F. 15 **G.** 30 **H.** 45 **J.** 60 **K.** 90

$5x + x = 6x = 180,\ x = 30$
Also, $y = 2x + x = 3x = 3(30) = 90$

Ans. K

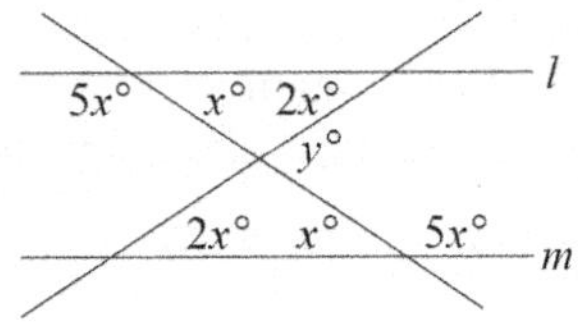

49. If the rabbit production in a farm increases in numbers by multiple of two in every 2 months, then how many months have elapsed to have the number of its production reaching 2^{11}?

A. 10 **B.** 20 **C.** 30 **D.** 40 **E.** 50

$2^1 \to 2^2 \to 2^3 \to 2^4 \to \ldots \to 2^{11}$
From 2^1 to 2^2, it takes 2 months.
From 2^1 to 2^3, it takes $(2 + 2)$ months.
From 2^1 to 2^4, it takes $(2 + 2 + 2)$ or (2×3) months.
……
$\therefore$ From 2^1 to 2^{11}, it takes (2×10) months = 20.

Ans. B

a	1	b
3	5	c
4	d	2

50. In the table of numbers above, the sum of the numbers in each row, column, and diagonal is the same. If the letters a, b, c, and d represent the integers 6 through 9, but not necessarily in that order, which letter represents the number 9?

F. a **G.** b **H.** c **J.** d
K. It cannot be determined from the information given.

$a = 8$	1	$b = 6$
3	5	$c = 7$
4	$d = 9$	2

Let's try with $a = 6$,
then the sum of the first column $= 6 + 3 + 4 = 13$.
$\therefore$ From the second col., $1 + 5 + d = 13 \quad \therefore d = 7$
From the first row, $a + 1 + b = 6 + 1 + b = 13 \quad \therefore b = 6$,
which contradicts $a = 6$. $\therefore a \neq 6$
Trying $a = 7$ in the same way, $a = 7$ does not work!
Let's try $a = 8$. Then, the column Sum $= 8 + 3 + 4 = 15$.
$\therefore$ For the second column, the sum $= 1 + 5 + d = 15 \quad \therefore d = 9$
Also, row sum, $a + 1 + b = 8 + 1 + b = 15 \quad \therefore b = 6$.
Also, $3 + 5 + c = 15 \quad \therefore c = 7$
$\therefore a = 8,\ b = 6,\ c = 7$ and $d = 9$

Ans. J

51. Set X has 6 members and set Y has 7 members. Set Z consists of all members that are in either set X or set Y with the exception of the k common members ($k > 0$). Which of the following represents the number of members in set Z?

A. $13 + k$ **B.** $13 - k$
C. $13 + 2k$ **D.** $13 - 2k$ **E.** 13

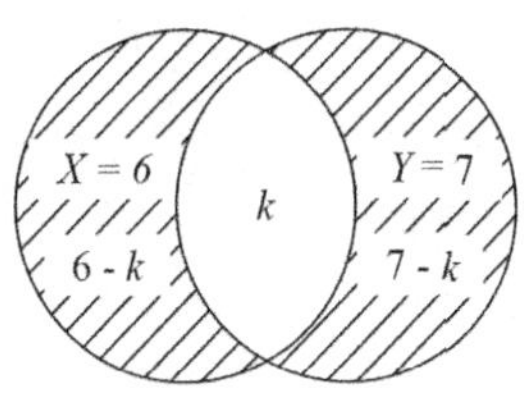

$Z = (X - k) + (Y - k)$
$= 6 + 7 - 2k = 13 - 2k$

Ans. D

52. If $0 < a < b$, then the solution to the equation $\frac{ax - b}{bx - a} = 2$ is $x =$

F. $\frac{2a - b}{2b - a}$ **G.** $\frac{2a + b}{2b - a}$ **H.** $\frac{a - b}{b - a}$ **J.** $\frac{2b - a}{2a - b}$ **K.** $\frac{2b + a}{2a + b}$

$\frac{ax - b}{bx - a} = 2$

$\therefore ax - b = 2(bx - a)$, or $ax - b = 2bx - 2a$

$\therefore 2a - b = 2bx - ax = (2b - a)x$

$\therefore x = \frac{2a - b}{2b - a}$

Ans. F

53. If a side of a cube is 1 foot, what is the volume of the cube, in cubic inches?

A. 1 **B.** 1728 **C.** 1440 **D.** 256 **E.** 625

The volume = 1 foot × 1 foot × 1 foot
= 12 in × 12 in × 12 in = 1728 in^3

Ans. B

54. Let the operations $\triangle$ and $\frown$ be defined for all real numbers a and b as follows.

$a \triangle b = 2a + 3b$

$a \frown b = 3a + 2b$

If $3 \triangle (4x) = (3x) \frown 4$, what is the value of x?

F. 1 **G.** $\frac{2}{3}$ **H.** $\frac{1}{3}$ **J.** 2 **K.** $\frac{1}{2}$

$3 \triangle (4x) = 2(3) + 3(4x) = 6 + 12x$

Also, $(3x) \frown 4 = 3(3x) + 2(4) = 9x + 8$

$\therefore 6 + 12x = 9x + 8$

$\therefore 3x = 2$

$\therefore x = \frac{2}{3}$

Ans. G

55. If $f(x) = \frac{1}{2}|x| + 3$, for which of the following values of x does $f(x) = f(-x)$?

A. -10 only
B. -10 and 10 only
C. All real x
D. All real x except 10
E. All real x except -10 and 10

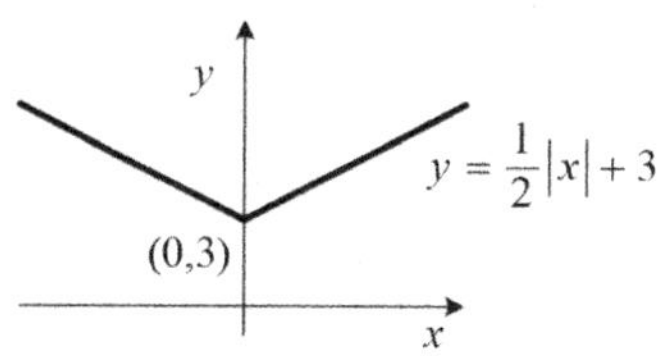

Since $f(x) = \frac{1}{2}|x| + 3$ is the even function, $f(x) = f(-x)$ is true for all real x-values.

Ans. C

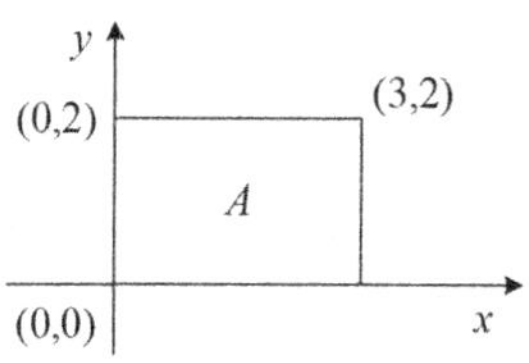

56. In the given figure, rectangle A contains all points (x,y). What is the area of a rectangle that contains all points $(2x, y - 1)$?

F. 12 **G.** 14 **H.** 18 **J.** 24 **K.** 48

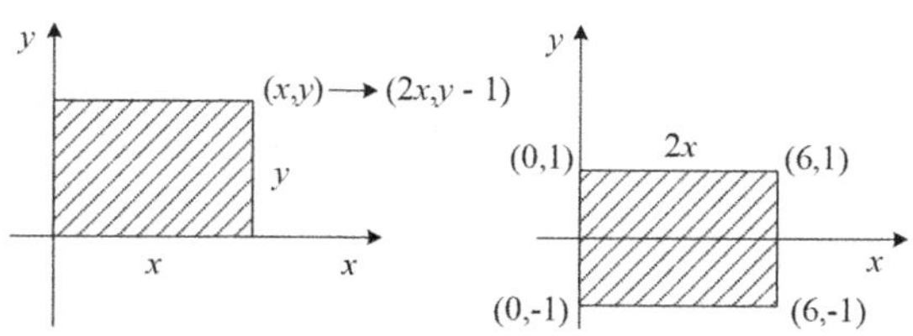

Since $x \rightarrow 2x$ has been twice stretched, while entire y has been shifted down 1 unit with its length unchanged, we get area, $A = 6 \times 2 = 12$

Ans. F

57. If $\sin\theta = m$ and $0 < \theta < 90°$, then $\tan\theta =$

A. $\frac{1}{m^2}$

B. $\frac{m}{\sqrt{1 - m^2}}$

C. $\frac{1 - m^2}{m}$

D. $\frac{m}{1 - m^2}$

E. $\frac{m^2}{\sqrt{1 - m^2}}$

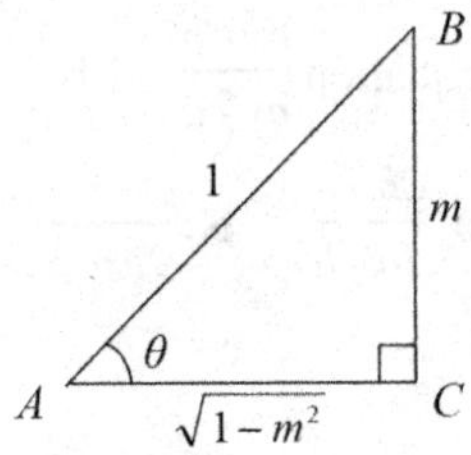

$\sin\theta = \frac{m}{1} = m$

$\therefore$ Using Pythagorean, $\overline{AC} = \sqrt{1 - m^2}$

$\therefore \tan\theta = \frac{m}{\sqrt{1 - m^2}}$

Ans. B

58. If $f(x) = 3\sqrt{x} + 1$, for all $x > 0$, then $f^{-1}(x) =$

F. $\frac{1}{9}(x - 1)^2$ **G.** $\frac{1}{3}(x + 2)^2$ **H.** $x^2 + 1$

J. $(x - 4)^2$ **K.** $(x + 2)^2 + 1$

Since inverse function is symmetric to $y = x$ line, and switches x into y and y into x, we get $y = 3\sqrt{x} + 1$ into $x = 3\sqrt{y} + 1$

$\therefore x - 1 = 3\sqrt{y}$ or $(\sqrt{y})^2 = \left(\frac{x - 1}{3}\right)^2 = \frac{1}{9}(x - 1)^2$

Ans. F

59. When $2x^2 + 6x + k$ is divided by $x - 2$, the remainder is 2. Which of the following is the value of k?

A. -18 **B.** -16 **C.** -10 **D.** 6 **E.** 14

Using Remainder Theorem, $P(2) = 2(2)^2 + 6(2) + k = 2$ or $20 + k = 2$

$\therefore k = -18$

Ans. A

60. y varies directly as the square of x. When $y = 6$, $x = 4$. If $y = 27$, then x could equal

F. $-6\sqrt{2}$
G. $-8\sqrt{2}$
H. $-10\sqrt{3}$
J. $-16\sqrt{2}$
K. $-64\sqrt{3}$

Since y varies directly as the square of x, we get $y = kx^2$.
But when $y = 6$, $x = 4$

$\therefore 6 = k(4)^2$

$\therefore k = \frac{6}{16} = \frac{3}{8}$

$\therefore y = \frac{3}{8}x^2$

Now, $y = 27 = \frac{3}{8}x^2$

$\therefore x^2 = \frac{27 \times 8}{3} = 72$

$\therefore x = \pm\sqrt{72} = \pm 6\sqrt{2}$

Ans. F

Made in the USA
Monee, IL
18 October 2026

66036816R00109